Instabilities in Dynamical Systems

NATO ADVANCED STUDY INSTITUTES SERIES

Proceedings of the Advanced Study Institute Programme, which aims
at the dissemination of advanced knowledge and
the formation of contacts among scientists from different countries

The series is published by an international board of publishers in conjunction
with NATO Scientific Affairs Division

A	Life Sciences	Plenum Publishing Corporation
B	Physics	London and New York
C	Mathematical and	D. Reidel Publishing Company
	Physical Sciences	Dordrecht, Boston and London
D	Behavioral and	Sijthoff International Publishing Company
	Social Sciences	Leiden
E	Applied Sciences	Noordhoff International Publishing
		Leiden

Series C – Mathematical and Physical Sciences

Volume 47 – Instabilities in Dynamical Systems

Instabilities
in Dynamical Systems

Applications to Celestial Mechanics

Proceedings of the NATO Advanced Study Institute
held at Cortina D'Ampezzo, Italy, July 30 - August 12, 1978

edited by

VICTOR G. SZEBEHELY

Dept. of Aerospace Engineering and Engineering Mechanics,
The University of Texas at Austin, Tex. 78712, U.S.A.

D. Reidel Publishing Company

Dordrecht : Holland / Boston : U.S.A. / London : England

Published in cooperation with NATO Scientific Affairs Division

Library of Congress Cataloging in Publication Data

NATO Advanced Study Institute, Cortina d'Ampezzo, Italy, 1978.
   Instabilities in Dynamical Systems.
   (Nato advanced study institute: Series C, Mathematical and physical sciences; v. 47)
   Includes index.
   1. Mechanics, Celestial—Congresses.   2. Stability—Congresses.   I. Szebehely, Victor G.,
1921–   II. Title.   III. Series.
QB349.N37   1978   523.01   79-11477
ISBN-13: 978-94-009-9425-6   e-ISBN-13: 978-94-009-9423-2
DOI: 10.1007/978-94-009-9423-2

Published by D. Reidel Publishing Company
P.O. Box 17, Dordrecht, Holland

Sold and distributed in the U.S.A., Canada, and Mexico
by D. Reidel Publishing Company, Inc.
Lincoln Building, 160 Old Derby Street, Hingham, Mass. 02043, U.S.A.

TABLE OF CONTENTS

PART IV: THE PROBLEM OF THREE BODIES

PART V: ABSTRACTS OF SEMINAR-CONTRIBUTIONS

PREFACE

As the director of the third NATO Advanced Study Institute
conducted in Cortina d'Ampezzo between July 30 and August 12,
1978 and as the editor of our Proceedings, it is my pleasure to
offer this volume to our colleagues.

The Institute took place, once again, at the Istituto
Antonelli where our "splendid isolation" was guaranteed by the
living-working-eating together arrangements. The frequent rains,
for which, of course, this director gladly accepted blame, increased the interest in the lectures and the few sunny days
allowed the establishment of friendly peer-relations among the
sixty-five amateur mountain-climbers. The highly professional
attitude of the participants in matters dynamical was sharply
contrasted with our dilettante approach to the Dolomites. So
we attacked and climbed the subject of instabilities in dynamics
"because it was there." It still is and we shall attempt to
climb it again.

I wish to extend my appreciation to the Scientific Affairs
Division of NATO, to the Italian Research Council and to The
University of Texas for their support. The faithful assistance
of the Director and staff of the Istituto Antonelli, the cooperation of all speakers and participants, the fatigueless diligence
of my co-director Professor G. Colombo and of my assistants
(Dr. G. Bianchini and Mr. R. McKenzie) deserve the highest
praise. The demanding and exacting job of typing the Proceedings
fell upon Miss B. Yates who met this challenge with her usual
patience and competence.

Cortina d'Ampezzo, Italy and   Victor Szebehely
Austin, Texas, USA   Director, NATO
                                Advanced Study Institute

Victor G. Szebehely (ed.), Instabilities in Dynamical Systems. ix.
Copyright © 1979 by D. Reidel Publishing Company.

LIST OF SPEAKERS AND PARTICIPANTS

Aarseth, S. J.   (Norwegian)   University of Cambridge, Institute of Astronomy, Madingley Road, Cambridge, U.K.

Antonopoulos, P.   (Greek)   Department of Astronomy, University of Patras, Patras, Greece

Auchmuty, J. F. G.   (U.S.A.)   Department of Mathematics, Indiana University, Swain Hall East, Bloomington, Indiana 47401 U.S.A.

Baumgarte, J.   (German)   Mechanik-Zentrum, Technische Universität Braunschweig Pockelsstr. 4, West Germany.

Baxa, P.   (Italian)   Universita Degli Studi di Trieste, Istitute di Geodesia e Geofisica Via dell Universita 7, Trieste, Italy.

Bianchini, G.   (Italian)   Department of Applied Mechanics, University of Padova, Via Venezia 1, 35100 Padova, Italy

Broucke, R.   (Belgian)   Department of Aerospace Engineering, University of Texas, Austin, Texas, 78712, U.S.A.

Bryant, J.   (U.S.A.)   Laboratoire de Mécanique Théorique, Faculté des Sciences et Techniques, Route de Gray, 25030 Besançon, Cedex, France

Carneiro, J.   (Portuguese)   Observ. "Prof. Manuel
                                              De Barros" Universidade
                                              Do Porto, Portugal

Carusi, A.   (Italian)   Laboratorio di Astro-
                                              fisica Spaziale, Reparto
                                              di Planetologia, V.le
                                              Regina Margherita 202,
                                              Roma, Italy

Colombo, G.   (Italian)   Istituto di Meccanica
                                              Applicata, Facoltà di
                                              Ingegneria, Università
                                              di Padova, V. Venezia 1,
                                              Padova 35100, Italy

Contopoulos, G.   (Greek)   Dept. of Astronomy, Uni-
                                              versity of Athens,
                                              Panepistimiopolis, Athens
                                              621, Greece

DeWitt-Morette, C.   (French)   Dept. of Astronomy,
                                              University of Texas,
                                              Austin, Texas 78712,
                                              U.S.A.

Dvorak, R.   (Austrian)   Astronomisches Institut,
                                              Universität Graz, Uni-
                                              versitätsplatz 5, A-8010
                                              Graz, Austria

Easton, R. W.   (U.S.A.)   Dept. of Mathematics,
                                              Univ. of Colorado,
                                              Boulder, Colorado 80309
                                              U.S.A.

Farinella, P.   (Italian)   Osservatorio Astronomico
                                              di Brera - 22055 Merate
                                              (Como), Italy

Forti, G.   (Italian)   Osservatorio Astrofisico
                                              di Arcetri, Largo E.
                                              Fermi 5, 50125 Firenze,
                                              Italy

Froeschlé, C.   (French)   Observatoire de Nice,
                                              BP 252, 06007 Nice,
                                              Cedex, France

Galgani, L.   (Italian)   Istituto di Fisica,
                                            Istituto di Matematica,
                                            Via Celoria 16, Milano,
                                            Italy

Garfinkel, B.   (U.S.A.)   Yale Univ. Observatory,
                                            Box 2023, Yale Station,
                                            New Haven, Connecticut,
                                            06520, U.S.A.

Goldreich, P.   (U.S.A.)   Page House, Cal. Tech.,
                                            Pasadena, California
                                            91126, U.S.A.

Gómez, G.   (Spanish)   Universidad Autonoma de
                                            Barcelona, Seccio de
                                            Matematiques, Bellaterra,
                                            Barcelona, Spain

Goudas, C. L.   (Greek)   Dept. of Mechanics,
                                            Univ. of Patras, Patras
                                            Greece

Gurel, O.   (Turkish)   I.B.M., 1133 Westchester
                                            Ave., White Plains, New
                                            York 10604, U.S.A.

Hadjidemetriou, J. D.   (Greek)   Dept. of Theoretical
                                            Mechanics, Univ. of
                                            Thessaloniki, Thessalo-
                                            niki, Greece

Henrard, J.   (Belgian)   Facultés Universitaires
                                            de Namur, 61 Rue de
                                            Bruxelles, 5000 Namur
                                            Belgium

Heppenheimer, T. A.   (U.S.A.)   Center for Space Science,
                                            11040 Blue Allium Ave.,
                                            Fountain Valley, Califor-
                                            nia 92708, U.S.A.

Hitzl, D. L.   (U.S.A.)   Lockheed Research Lab.,
                                            3251 Hanover Street,
                                            Dept. 52-56, Bldg. 201,
                                            Palo Alto, California
                                            94304, U.S.A.

Hubbard, M.	(U.S.A.)	Dept. of Mechanical Engineering, Univ. of California, Davis, California 95616, U.S.A.
Junkins, J. L.	(U.S.A.)	Dept. of Engineering Science and Mechanics, Virginia Polytechnic Institute and State Univ., Blacksburg, Virginia 24061, U.S.A.
Katsiaris, G. A.	(Greek)	Dept. of Mechanics, Univ. of Patras, Patras, Greece
Kinoshita, H.	(Japanese)	Tokyo Astronomical Observatory, 2-21-1 Osawa, Mitaka Tokyo, Japan
Llibre, J.	(Spanish)	Universidad Autonoma de Barcelona, Seccio de Matematiques, Bellaterra, Barcelona, Spain
McKenzie, R.	(U.S.A.)	Dept. of Aerospace Engineering, Univ. of Texas, Austin, Texas 78712, U.S.A.
Magnenat, P.	(Swiss)	Geneva Observatory, 1290 Sauverny, Switzerland
Markellos, V.	(Greek)	Dept. of Astronomy, Univ. of Glasgow, Glasgow 512 8QQ, Scotland, U.K.
Martinez, J.	(Spanish)	Catedra Especial de Technologias del Espacio, Universidad Politecnica de Barcelona, Spain
Martins, R. V.	(Brazilian)	97, Rue de l'Amiral Mouchez, 75013 Paris, France
Message, P. J.	(U.K.)	Liverpool Univ., Dept. of Applied Mathematics and Theoretical Physics, Liverpool L 69 3BX, U.K.

Mignard, F.	(French)	C.E.R.G.A., 8 bd. E. Zola, 06130 Grasse, France
Milani, A.	(Italian)	Institute of Mathematics, Univ. of Pisa, Via Derna 1, 56100 Pisa, Italy
Moons, M.	(Belgian)	Facultés Universitaires Notre Dame de la Paix, Dept. de Matématiques, Rue de Bruxelles 61, 5000 Namur, Belgium
Morse, E.	(U.S.A.)	L-439 Lawrence Livermore Lab, Livermore, California 94566, U.S.A.
Nahon, F.	(French)	Laboratoire de Dynamique et Statitique Stellaire, 11 Rue Pierre et Marie Curie, 75231 Paris, Cedex, France
Nobili, A.	(Italian)	Università di Pisa, Istituto di Scienze dell'Informazione, Via di Gello 60, 56100 Pisa, Italy
Puel, F.	(Franch)	Observatoire de Besançon, 25000 Besançon, France
Rapaport, M.	(French)	Observatoire de l'Université de Bordeaux, 33270 Floirac, France
Richardson, D. L.	(U.S.A.)	Dept. of Eng. Science, Univ. of Cincinnati, Cincinnati, Ohio 45221 U.S.A.
Robin, I.	(U.K.)	Dept. of Astronomy, Univ. of Glasgow, Glasgow, G 12 8QQ, Glasgow, Scotland, U.K.
Roy, A. E.	(U.K.)	Dept. of Astronomy, Univ. of Glasgow, Glasgow G 12 8QQ, Glasgow, Scotland, U.K.

Rüssmann, H.   (German)   Fachberreich Mathematik
                                        der Universitat, Postfach
                                        3980, 6500 Mainz, West
                                        Germany

Scholl, H.   (German)   Astronomisches Rechen-
                                        Institut, Heidelberg,
                                        West Germany

Sessin, W.   (Brazilian)   Departemento de Astro-
                                        nomia, CTA-ITA Sao Jose
                                        dos Campos, Sao Paulo,
                                        Brazil

Simó, C.   (Spanish)   Universidad Autonoma de
                                        Barcelona, Secciò de
                                        Matematiques, Bellaterra,
                                        Barcelona, Spain

Standaert, D.   (Belgian)   Facultés Universitaires
                                        Notre Dame de la Paix,
                                        Rempart de la Vierge,
                                        8 B-5000 Namur, Belgium

Stey, G. C.   (U.S.A.)   Instituts Internationaux
                                        de Physique et de Chimie,
                                        fondés par E. Solvay,
                                        U.L.B., Chimie-Physique
                                        2, 1050 Bruxelles, Belgium

Szebehely, V.   (U.S.A.)   Dept. of Aerospace Engi-
                                        neering, Univ. of Texas,
                                        Austin, Texas 78712 U.S.A.

Tremaine, S.   (Canadian)   Institute of Astronomy,
                                        Madingley Road, Cambridge
                                        CB 3 OMA, England

Vazquez, L.   (Spanish)   Departemento de Fisica
                                        Teorica, Facultad de
                                        Ciencias Fisicas, Univer-
                                        sidad Complutense,
                                        Madrid 3, Spain

Vicente, R.   (Portuguese)   R. Mestre Aviz 30, R/C
                                        Lisboa, Portugal

Vinti, J.   (U.S.A.)   Bldg. W 91 - 202 M.I.T.
                                        Cambridge, Massachusetts
                                        02139, U.S.A.

Waldvogel, J.   (Swiss)   ETH-Zurich, Seminar fur
                                        Ange. Mathematik, 8092
                                        Zurich, Switzerland

Yokoyama, T.   (Brazilian)   Instituto Astronomico
                                        E Geofisico USP, Depar-
                                        temento de Astronomia,
                                        Caixa Postal 30627 –
                                        CEP 01000, Sao Paulo,
                                        Brazil

Zappalà, V.   (Italian)   Osservatorio Astronomico
                                        di Torino, 10025 Pino
                                        Torinese (To), Italy

REMARKS BY DR. BORIS GARFINKEL ON THE OCCASION

OF THE BIRTHDAY OF PROFESSOR VICTOR SZEBEHELY

I am glad to be here on this ceremonial occasion to wish a happy birthday to Victor Szebehely, our Director and friend. It is most fitting that we use the opportunity to pay special tribute to the man who has done so much to promote the cause of Celestial Mechanics throughout the years. By bringing together the leading scholars of Europe and America he has performed a unique public service. How he was able to accomplish this remarkable feat in these days of severe financial stringency, to me is a mystery, undoubtedly involving some uncanny combination of personal charm and some devious diplomacy in the grand tradition of the old Austro-Hungarian Empire. We are particularly grateful to him that again we meet in the beautiful Cortina d'Ampezzo, enjoying the fine hospitality of the Istituto Antonelli. Here the magnificent panorama of lofty mountains provide a fitting background to the pinnacles of human thought built by the enduring works of Newton, Lagrange, Euler, Hamilton, Jacobi and their modern successors.

May you, Victor, continue to enjoy good health and many more years of productive research, and may there be many more Cortinas in our future.

Victor G. Szebehely (ed.), Instabilities in Dynamical Systems. xix-xxii.
Copyright © 1979 by D. Reidel Publishing Company.

REMARKS BY PROFESSOR A.E. ROY ON THE OCCASION

OF THE BIRTHDAY OF PROFESSOR VICTOR SZEBEHELY

When we come to Cortina, we are immediately aware of the
majesty of the surroundings, in particular the Dolomite Mountains.
All of us wish to climb and explore them; some of us achieve
their goal. In doing so we not only learn about the landscape
we toil over, but also learn of our own capabilities. Achievement and struggle brings understanding and satisfaction. From
such peaks we also get a better view of the landscape. In the
landscape of Celestial Mechanics there are also mountains to explore. Among them are Mount Newton, Mount Jacobi, Mount Poincare
and so on. They are peaks of achievement, landmarks and sign-
posts in the terrain we wish to cross. One of these peaks is
justly named Mount Victor Szebehely for by a study of his work,
his paper, his celebrated book on the three-body problem, many
of us here have been aided in our own attempts to make some contribution to the exploration of the landscape of Celestial Mechanics.

Victor Szebehely has achieved a world-wide reputation in
our subject. But ladies and gentlemen, he has also acquired a
reputation as an expert in at least one other field. Let us tell
you how I discovered this.

Some years ago I was travelling with a small party of peo-
ple in Greece and one night we had dinner with some friends in
the Rion Hotel at Patras. Victor Szebehely was at the dinner
table. I noticed throughout the meal how very interested a
young lady member of my party was in Victor's conversation. At
breakfast the following morning she asked us: "Wasn't Dr.
Szebehely a fascinating man?"

I hastened to agree or --- memory fails me regarding my
precise words --- said something to the effect that I couldn't
say I had noticed!

"What does he do?", she inquired.

"He is a world authority on the three-body problem!"

"Goodness! What is that?"

I thought rapidly.

"Roughly speaking, it is concerned with the possible actions of three bodies. You know: two is company - three's a crowd (deux peuple est joie: trois est une foule).

"And Professor Szebehely is an expert in <u>that</u>?"

"Yes, I replied. "In his studies he has classified all the actions of three bodies -- capture...interplay...escape...!

Her eyes widened.

"Good gracious me, an expert in the most intimate human relationships. If I'd known last night I could have asked his advice on some of my problems!"

It is not given to many of us, Victor, to be international authorities on fields as diverse as Celestial Mechanics and human relations.

But we talked of mountains as landmarks in a landscape.

Birthdays are also landmarks at which we take stock and survey the landscape of our lives. When we were young, birthdays were events we eagerly anticipated. This is not the case as we grow older. Not only do we suspect that the passage of time is accelerating but we tend to look at our past from the standpoint of a birthday and feel that we should have used the past year to much better effect. But we are only as old as we feel and, judging by Victor's activities - research, teaching, organizing NATO Advanced Study Institutes, etc., etc., it is obvious to me that he should be regarded as one of the few genuine pre-war teenagers left in circulation.

I think it was the astronomer Dr. Abbot who, having officially retired at the age of 65, continued to work fruitfully until his 100th birthday and who said in answer to his friends' congratulations: "Thank you, gentlemen, and do you know, I am also encouraged by the fact that at the beginning of my second century I find myself much steadier on my feet than at the beginning of my first!"

I am sure that Victor Szebehely's future years will be at least as fruitful in his various activities as in the past. And so, ladies and gentlemen, friends and colleagues of Victor Szebehely, I ask you to lift your glasses on the occasion of this happy celebration of our Director's birthday and join me in saying: "Victor, very many happy returns of the day!"

GREETINGS BY DR. P. J. MESSAGE

I wish to recall how the late Dr. Dirk Brouwer, amongst his many contributions to the well-being of celestial mechanics, had done much to organize meetings from time to time for many of the workers in the field to exchange ideas and to stimulate research effort. Victor Szebehely has done much to continue this tradition, and undertaken the responsibility and labor of arranging many meetings such as this at opportune times and places. He has left us very much indebted to him.

INTRODUCTION

Following the principle according to which not much is
accomplished by working on easy problems - even if one succeeds
in solving them - we selected one of the prize-problems in dynamics, stability. The kind reader will judge how much was
accomplished in the tutorial treatment of the group of problems
associated with stability - or, indeed, concerning the philosophy of stability research.

Our title refers to instabilities and aptly so since to
predict instabilities may be the first step in research on stability and it is often more important. Examples abound: numerical experiments often show the sudden appearance of instabilities;
linear instabilities may be extended to nonlinear results; the
Hill-Lyapunov method is designed directly for predicting possible instabilities, etc.

This volume and the lectures emphasize the tutorial aspects
of the subject with Seneca's motto: Docendo discimus. We at-
tempted to discipline the great variety of definitions of stability used often in a cavalier fashion in the literature. The
speakers were asked to follow only a few rules: speak loudly
and slowly, treat non-integrable dynamical systems and define
their own concepts of stability. The seminar sessions (summarized by the Abstracts in this volume) were equally demanding as
were the organized discussions on the present state of our
knowledge concerning the stability of the solar system, on the
question of stability of periodic orbits in the general problem
of three bodies and on non-integrable dynamical systems.

The unresolved question of the stability of the solar
system was approached by a variety of methods. It may be concluded that stability is suspected but not proven for more than
10^6 years. Four-body periodic planetary orbits often show capabilities to behave in a resonance-induced unstable manner,
while various three-body combinations (such as the Sun, Jupiter
and Saturn) show stability according to Hill's definition. Our

Victor G. Szebehely (ed.), Instabilities in Dynamical Systems, xxiii-xxiv.
Copyright © 1979 by D. Reidel Publishing Company.

approaches at present often neglect accretion and tidal effects which must have been essential (and still might be) in the development of the solar system. Formal proof of the invariability of the semi-major axes to any order, using Lie-series-transformations is indeed reassuring as it was originally to Laplace, but Poincaré's sobering warning awoke us to the rude reality of no-convergence. The discussion of the celebrated K-A-M theorem established once again the known difficulties of its application to the solar system: satisfaction of the irrationality conditions on the mean motions is undecided since the frequencies cannot be measured with sufficient accuracy; the condition set in the theorem on the smallness of the perturbations are much too restrictive for applications to the solar system; the dimensionality of the solar system is too high; nongravitational effects may not be neglected, etc.

Stability of periodic orbits in the three-body problem was subjected to vigorous scruting with the raison d'être that (i) besides a few other special orbits these seem to be the only solutions known for arbitrarily long times, (ii) the (linear) stability of these orbits may be determined with relative ease, and (iii) Poincaré's dictum suggests that these orbits offer the only "breach" through which one may comprehend non-integrable dynamical systems such as the problem of three bodies.

An impromptu and exciting discussion revealed the considerable confusion existing about the definition of non-integrable dynamical systems in principle and in practice. Participants were urged to keep this subject alive via correspondence.

The applications covered cluster formation and gravitational instabilities in the universe, stability of planetary and satellite systems, unstable exchanges in triple stellar configurations and a variety of perturbed oscillators. The highly controversial subject of the effect of resonance on the stability of non-linear (and non-integrable) systems was bravely encountered and the formation, discovery and dynamics of planetary rings and of the asteroid belt were treated with the certainly deserving, considerable care.

Stabilization of numerical integration methods, bifurcation theory and the possible interrelation between statistical mechanics and modern dynamics were discussed in detail.

At one point during a discussion session one of the participants with an easily recognizable French accent proposed that we should conduct an Institute in the future on dynamical astronomy, where all references to Poincaré and to periodic orbits are forbidden. And with this, admittedly sacrilegious note, this Introduction is concluded.

Cortina d'Ampezzo, Italy and   Victor Szebehely
Austin, Texas, USA   Editor of the Proceedings

PART I

FUNDAMENTAL CONSIDERATIONS OF STABILITY

SIMPLE NON-INTEGRABLE SYSTEMS WITH TWO DEGREES OF FREEDOM

R. Broucke

The University of Texas at Austin

1.   INTRODUCTION

The present text gives some numerical results of our research on the solutions of simple, non-integrable dynamical systems.

The numerical results belong to four dynamical systems with two degrees of freedom. These are some of the simplest non-integrable dynamical systems that can be found, at least when expressed in rectangular coordinates. The aspects that are des- cribed center around the periodic solutions and their stability for the first three systems, and escape solutions for the last system.

The first system that was studied in some detail is the well known Henon system (Hénon-Heiles, 1964). We show the existence of five families of very simple periodic solutions. Two families are rather remarkable because of the existence of a bifurcation between these families. The periods are the same on both fami- lies, at the bifurcation, as well as the stability indices which are equal to +2 (Whittaker, 1960, page 404). The four eigen- values of the monodromy matrix are thus all equal to +1 (Wintner, 1947, page 104).

The second system has different equations of motion (Conto- poulos, 1965) but is fairly similar to the first system. Most of the periodic solutions are similar to those of the first system. In particular, we show that the same bifurcation between families of periodic solutions that was discovered in the Henon system also exists in the Contopoulos system.

3

Victor G. Szebehely (ed.), Instabilities in Dynamical Systems. 3-24.
Copyright © 1979 by D. Reidel Publishing Company.

The third system was obtained by modifying the equations of motion of the Henon system in such a way as to obtain a homogeneous potential. This homogeneity has many consequences for the behavior of the system. In particular, we show the existence of the eleventh integral (Losco, 1977). We also show that all the periodic solutions of a given family differ only by a scale factor and have thus all the same stability index. This fact makes the type of bifurcations that we found in the first two systems impossible in the present system.

The last system was included in the present article in order to show an example of a nonlinear system which has only escape solutions and no periodic solutions.

The other aspects that are being studied at the moment deal with the classification of chaotic and quasi-periodic solutions, the existence of new uniform integrals and the distinction between integrable and non-integrable systems.

2.   REMARKS ON THE INTEGRABILITY OF A DYNAMICAL SYSTEM

We are mostly concerned with Lagrangian or Hamiltonian dynamical systems with n degrees of freedom. These systems are by far the most important and best known. However, we first make a remark about non-Hamiltonian systems. In general, a dynamical system with n degrees of freedom is represented by a system of differential equations of order 2n:

$$\ddot{q} = f(q,\dot{q},t),$$

which can be reduced to a system of the form

$$\dot{x} = f(x,t),$$

where q is an n-vector and x a 2n-vector. All that is needed to solve the last system in the neighborhood of an initial point x_o is essentially the Lipschitz condition. The solution is then of the form

$$x = x(x_o,t),$$

and is valid in some neighborhood of x_o. These 2n equations can, in principle, be solved for the constants x_o:

$$x_o = g(x,t)$$

and it is seen that 2n integrals are found. In general, 2n integrals are required to solve a <u>general</u> dynamical system with n degrees of freedom (at least one of these integrals contains the time), in contrast with a <u>Hamiltonian</u> dynamical system that can be considered solved when n integrals are known. Another important point about the 2n integrals given above is that very often they are only <u>local</u> integrals, valid in a small neighborhood of values for the constants x_o. The integrals that we usually desire are the <u>global</u> uniform integrals, valid in the complete space.

We consider a Hamiltonian system with n degrees of freedom as integrable if it is possible to find n independent global uniform integrals of the system. (Remark: m integrals are said to be independent if the matrix formed with the gradients of these m integrals is of rank m.) This will, in general, be done by successive canonical transformations. A system is nonintegrable if it has less than n global integrals. Note that a non-integrable system may have one or several local uniform integrals valid only in subsets of the phase space.

The previous definition of an integrable system is the one which is accepted by most classical authors. In particular, it is in agreement with Wintner (1947, page 144). However, even Wintner is not completely certain what an integrable system is. He uses the expression, "<u>One is inclined</u> to consider a system as integrable if (but not only if) it can be split into n systems with a single degree of freedom, by means of explicit transformations of the coordinates and the time variable." Wintner's remarks clearly indicate that by integrability he means "reducible to quadratures" (page 143). But he then goes on and says that it is "quite undefined what an integrable system is." He also makes a remark that "along the line of Poincaré's dictum, a system can be neither integrable nor nonintegrable, but more or less integrable."

I believe that this last statement of Wintner can be understood in the light of modern research in dynamics. There seems to be quite a smooth transition of properties from the integrable to the nonintegrable system, rather than a discontinuous distinction.

I also feel that it is somewhat curious that Wintner does not mention Poincaré's non-existence theorem for the additional integrals, although Poincaré's work on this subject (1892) is considered an important cornerstone in the theory of integrals of dynamical systems. However, as has been pointed out by many authors (Cherry, 1924), there are serious restrictions to Poincaré's theorem on the nonexistence of additional integrals of a system. Essentially, one can say that Poincaré's theorem only applies to Hamiltonians which are expansions in a small parameter

$$H = H_o + \mu H_1 + \mu^2 H_2 + \cdots$$

and integrals which are similar expansions in μ. In other words, Poincaré considers dynamical systems which are perturbations of an integrable system Hamiltonian H_o. The zero-order approxim- ation H_o contains only half the canonical variables (moments, for instance) which are therefore constants of the motion and the H_o-problem is completely integrable. The integrals that Poincaré is looking for are Fourier series in the n coordinates, with the coefficients functions of the n momenta. (Poincaré's proof is reproduced by E. T. Whittaker in his "Analytical Dynam- ics," section 165, page 380.)

Another imporant contribution on the theory of integrals that must be mentioned is by Whittaker (1960, Chapter 16). He also expands the Hamiltonian in the square roots of the momenta and Fourier series of the coordinates. This allows him to dis- cover the so-called adelphic integral. This work is also closely related to the so-called Birkhoff normalization of a dynamical system. This normalization is a series of canonical transform- ations which leads to an integrable asymptotic approximation of the system. The new Hamiltonian is essentially a series expan- sion in the momenta only. These momenta are thus constants and provide the n integrals of the problem. If the problem is not integrable, the number of terms is infinite as well in the nor- malized Hamiltonian as in the expression for the fundamental frequencies (which are the partial derivatives of the Hamiltonian with respect to the momenta).

Finally, let us indicate the possibility of making numerical tests on a digital computer, in order to obtain information on the existence of integrals. These tests are relatively easy with two-degree-of-freedom systems. First of all, the method of surfaces of section is of great value; the invariant curves (level curves) are smooth curves if a second uniform integral exists. These curves break down in islands and finally in clouds of random points when the integral ceases to exist. If the inte- gral exists and is global (rather than local), the problem may be considered integrable.

Another possible numerical tests consists of studying the characteristic exponents of the periodic solutions of the system. In a Hamiltonian system, there is usually a pair of zero charac- teristic exponents for each integral of the system. Such a test is usually easy to make. The interpretation of the results may be complicated, however, because of several reasons: there may be ordinary and singular periodic solutions (Whittaker, 1960, pp 395 and 406); it may also happen that on some families of periodic solutions, the two integrals are not independent (the gradients parallel), in which case a pair of non-zero characteristic exponents appears.

3.   THE SIMPLE DYNAMICAL SYSTEM ONE (SDS1).

This is essentially the system that was studied by Henon and Heiles (1964) and then by many other researchers. The system has the conservative potential function

$$U = \frac{-1}{2}\,(x^2 + y^2) - \alpha(xy^2 - \frac{x^3}{3}) = \frac{-r^2}{2} + \alpha\,\frac{r^3}{3}\,\cos3\theta.$$

Note that this system differs from the Henon–Heiles system in two minor points: first of all the coordinates x and y have been interchanged in order to obtain here a system which is symmetric with respect to the x-axis. Secondly, a constant parameter α has been introduced. If this parameter is small we are in the presence of a perturbed harmonic oscillator with two dimensions. However, most of our investigations correspond to the numerical value $\alpha = +1$.

The problem has three symmetry axes, one of them being the x-axis, and four equilibrium points. The only stable equilibrium point is at the origin, where the energy is a minimum. The other three unstable equilibrium points are at the vertices of an equilateral triangle, at $(1/\alpha,0)$ and $(-1/2\alpha$, $\pm\sqrt{3/2\alpha})$.

We notice, by setting y=0 in the equations of motion of the Henon system, that in it is imbedded a one-degree-of-freedom system

$$\ddot{x} = -x + \alpha x^2 = \frac{\partial}{\partial x}\,(-\frac{x^2}{2} + \alpha\,\frac{x^2}{3})$$

which represents a nonlinear oscillation on the x-axis. An independent study of the solutions of this system will give some information about the original system. In particular, it is interesting to study the two-dimensional stability (in the y-direction!) of the periodic solutions of the x-system.

We have numerically computed the most important periodic solutions of the SDS1. We have restricted ourselves to those periodic solutions which are symmetric with respect to the x-axis. Therefore, these solutions have two orthogonal crossings with the x-axis, one being the initial point $(x_o,0,0,\dot{y}_o)$ and the other the final point of the numerical integration. Only a half revolution has to be integrated in the case of symmetric solutions. We present here only the so-called <u>one-arc</u> periodic solutions; each half revolution consists of <u>one</u> single arc from the initial orthogonal crossing to the final orthogonal crossing corresponding to t = T/2.

We found that there are essentially <u>five</u> families of one-arc symmetric periodic solutions with the numerical value α = 1. All our numerical calculations are limited to α = 1 for this system. The families have been designated by the letters A,B,C, D,E, in the order of discovery. The two families A and C originate at the equilibrium point (0,0) while the family E originates at the equilibrium point (1,0). The two families B and D do not seem to be directly related to the equilibrium points of the system. The families A, B and E are oscillations, while C and D are of the librations type. The initial conditions diagram (x_o, $\dot{y}_o$) for the five families is shown in figure 1. Note that we only show the upper half of the diagram corresponding to $\dot{y}_o > 0$. The lower half is identical because the problem is reversible.

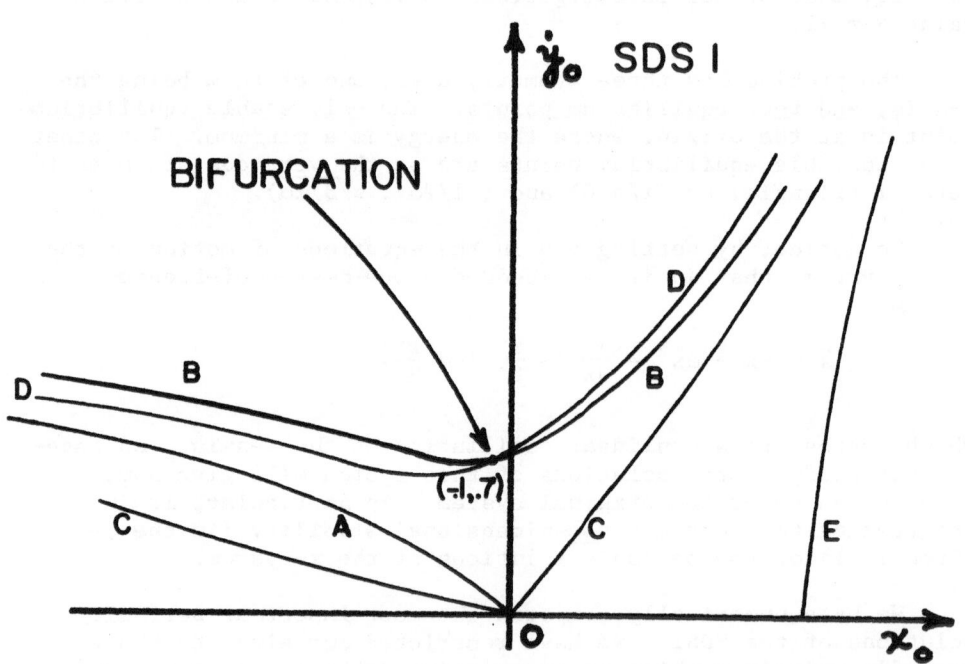

Figure 1. System 1; Initial conditions of periodic solutions.

In the neighborhood of the origin, the family A starts with vertical oscillations which are solutions of the linearized system: $x = 0$, $y = \varepsilon \sin t$, where ε is the parameter of the family (the amplitude). In the neighborhood of the origin, the half-period is ε. When the value of x_o decreases (x_o is < 0 over the whole family) the amplitude becomes larger and both the period and the energy start increasing. Some of the solutions are shown in figure 2.

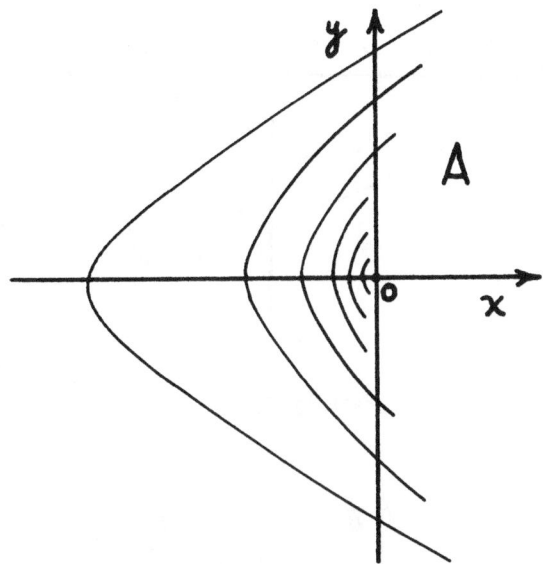

Figure 2. System 1; Family A of periodic solutions.

The family C also starts in the neighborhood of the equilibrium point at the origin. In the limit, the solutions are circles in the x-y plane: $x = \varepsilon \cos t$, $y = \varepsilon \sin t$. When the size increases, the shape of the orbits becomes triangular, while the energy increases (starting at 0). The general form of the orbits is shown in figure 3.

The family E starts at the equilibrium point (1,0) with oscillations of the form $x = 1$, $y = \varepsilon \sin \beta t$, with $\beta = \pi/1.8138$ (i.e., with the half-period 1.8138). The energy starts at 1/6 and increases. The half-period decreases continuously when the size increases. Figure 4 shows the shape of a few solutions.

The family B is another family of oscillations. It does not start at any one of the equilibrium points. The minimum value of $\dot{y}_o$ in this family is reached at about $x_o = -0.1729$, $\dot{y}_o = 0.7007$. A remarkable property of this family is the bifurcation point with the family D. At this point the stability index k is +2 and the energy of the family is a minimum (E = 0.2529). The

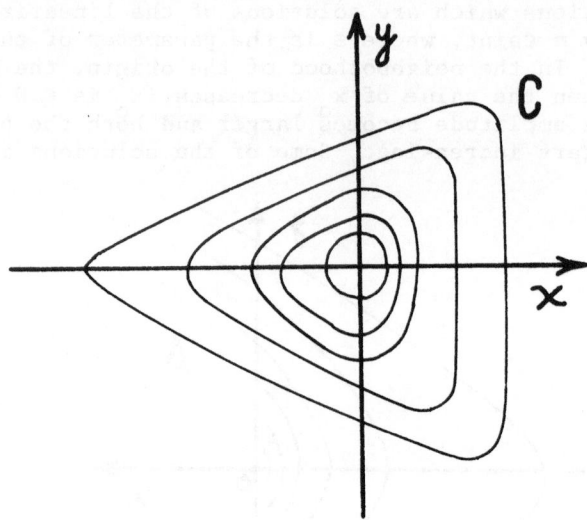

Figure 3. System 1; Family C of periodic solutions.

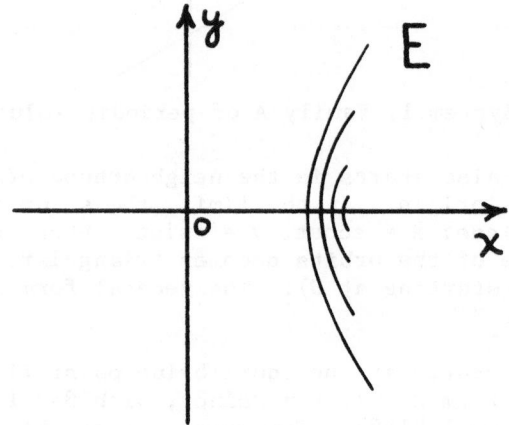

Figure 4. System 1; Family E of periodic solutions.

solution starts at $x_o = -0.070$, $\dot{y}_o = 0.707$, which is thus the intersection point of the two families. In figure 5 we show three solutions of family B, the solution in the middle being the bifurcation solution with family D.

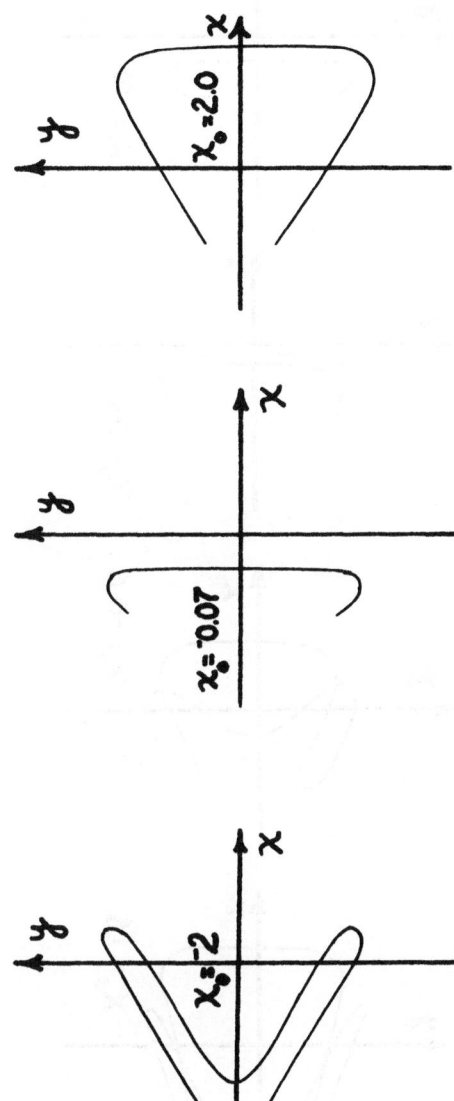

Figure 5.   System 1; Family B of periodic solutions.

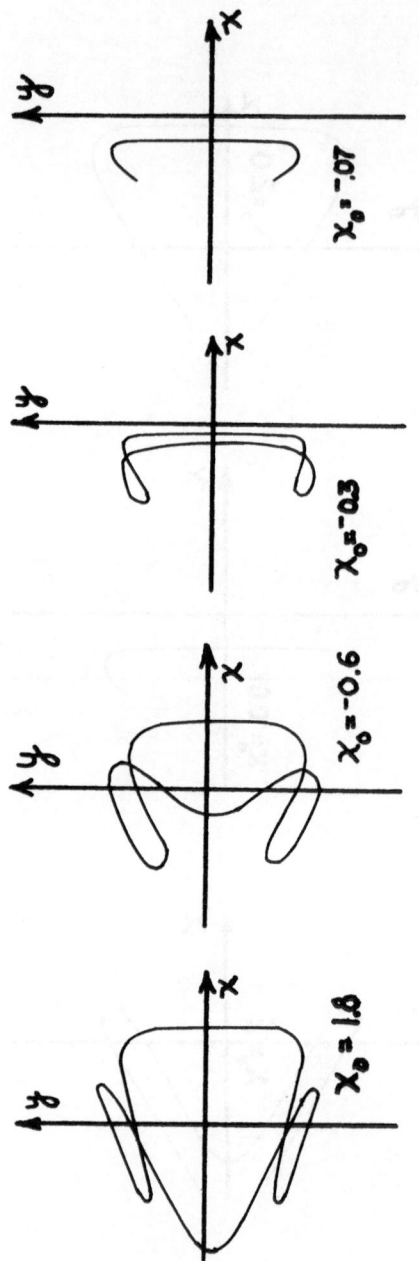

Figure 6. System 1; Family D of periodic solutions.

The family D of circulations can be considered as originating as a bifurcation from the family B at the point $(x_o, \dot{y}_o)$ = (-0.07, 0.70). The periods of the two families are the same near this point. However, the solutions of family D form a closed loop in the (x,y)-plane (while in family B we have a path with zero velocity points at the two end-points). The stability study and the special values of the characteristic exponents of these solutions in the neighborhood of the point (-.070, 0.70) support the theory of the presence of a bifurcation between two families at this point (see Figure 6, previous page). More details on the stability of the orbits of the Henon system are given in section seven below.

4.   THE SIMPLE DYNAMICAL SYSTEM TWO (SDS2).

One of the most simple nonintegrable dynamical systems that can be found is the system with the following potential function:

$$U = -\frac{1}{2}(x^2 + y^2) - \alpha xy^2.$$

This is a system that was extensively studied by Contopoulos (1965), except for some minor changes of notations. For instance, we introduced the parameter α in order to be able to study the transition from the integrable harmonic oscillator to the nonintegrable perturbed problem.

The system has three equilibrium points, (0,0) and $(-1/2\alpha, \pm \sqrt{2}/2\alpha)$, of which only the first one is stable.

We have computed the one-arc periodic solutions of the Contopoulos system, with the numerical value $\alpha = +1$. We have again restricted ourselves to periodic solutions which are symmetric with respect to the x-axis. Therefore, the initial conditions are still of the form $(x_o, 0, 0, \dot{y}_o)$.

The principal conclusion is that four families of periodic solutions are very similar to those of SDS1 (A,B,C,D) while on the other hand the family E does not exist for the system SDS2. This is to be expected because SDS2 does not have the equilibrium point at (1,0). The similarity of the $x_o, \dot{y}_o$-graphs is also evident (Fig. 7). For the families A,B and D, all the solutions look exactly similar to the solutions of the same families of SDS1 and for this reason we do not reproduce them. In particular, the bifurcation between families B and D is again present here, at the approximate point $(x_o = -0.10, \dot{y}_o = +0.645)$. We notice that, at the branch point, the family D goes through a minimum energy (0.213) and minimum period (7.4013). At this point the family B has minimum energy but no extremum for the period.

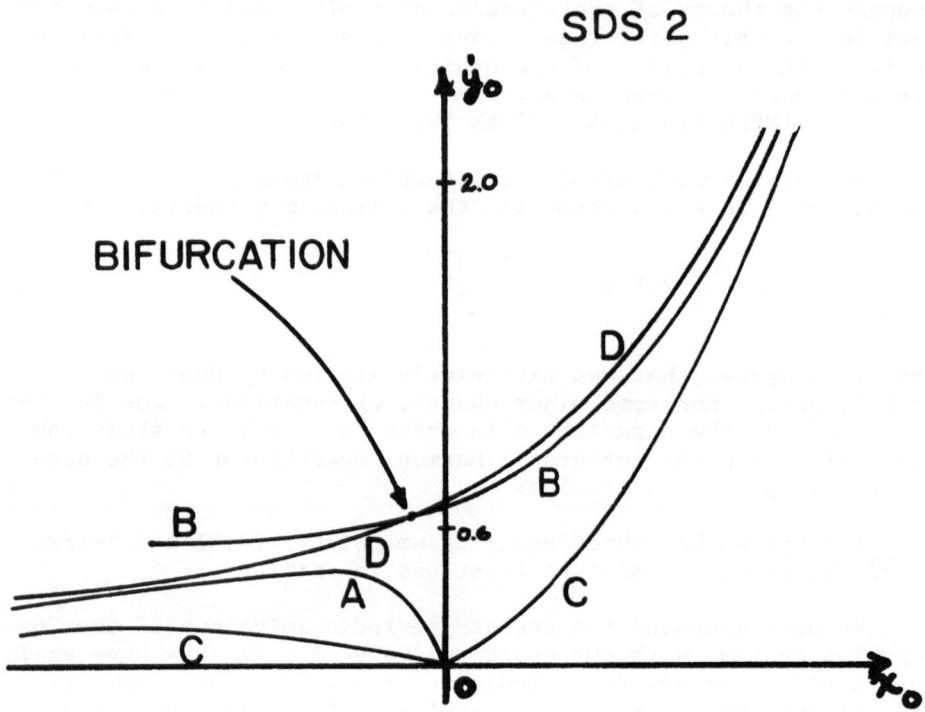

Figure 7. System 2; Initial conditions of periodic solutions.

        We reproduce in Figure 8 a few of the orbits of family C,
this being the only family which displays a notable difference in
appearance when compared with the corresponding family of SDS1.
We notice that near the origin the orbits are nearly elliptic,
with one axis twice the length of the other.

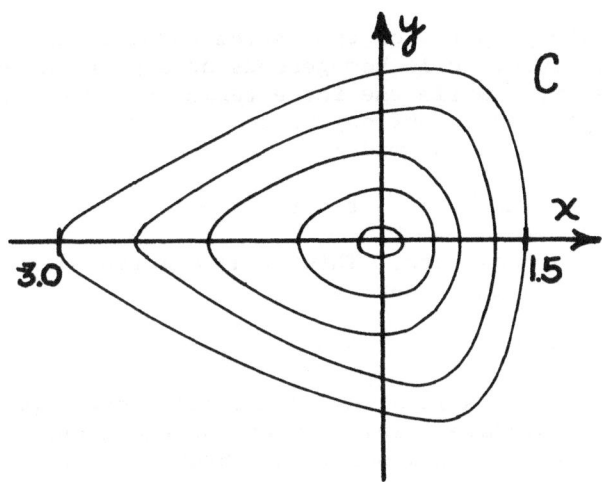

Figure 8.   System 2; Family C of periodic solutions.

5.   THE SIMPLE DYNAMICAL SYSTEM THREE (SDS3).

This system has been investigated because it is <u>not</u> a perturbation of a harmonic oscillator in two dimensions similar to most nonlinear problems found in the literature. This new system, which is the Henon system, where the lowest order terms have been dropped, will behave in the same way as the Henon system when x and y are large, but the behavior is expected to be totally different near the origin. In other words, the comparison of this system with the Henon system (SDS1) will show us the effects of the harmonic terms $(x^2 + y^2)/2$ in the potential.

The potential function of our system three is:

$$U = -\alpha(xy^2 - \frac{x^2}{2}).$$

Note that the origin here is the only equilibrium point and it is of the unstable type.

We could say that this removal of the harmonic terms $(x^2 + y^2)/2$ could also be done with the system two (Contopoulos) as well as the system one (Henon). We have effectively done this

with system two. The result is the rather unusual system (SDS4)
described in the next section.

A remarkable property of the system three is that it is of
the homogeneous type: U is homogeneous of degree three in the
two coordinates x,y, while the force terms in the differential
equations are homogeneous of degree two. Therefore, the following change of scale can be made:

$$x \rightarrow x\lambda \quad ; \quad t \rightarrow t\beta,$$

where λ and β are constants. This transformation leaves the
equations of motion invariant if

$$\lambda\beta^2 = 1 \quad ; \quad \beta = \lambda^{-1/2}.$$

The homogeneity results in a few additional properties, as we
shall see. We designate a solution of the equations of motion
by $q(t) = x(t),y(t)$. A new solution, obtained by change of scale
is then

$$\lambda q(\lambda^{1/2}t) \quad ; \quad \lambda^{3/2}\dot{q}(\lambda^{1/2}t).$$

By taking the partial derivative in λ and next setting $\lambda = 1$, we
obtain a solution of the variational equations

$$q + \frac{1}{2} t\dot{q} \quad ; \quad \frac{3}{2} \dot{q} + \frac{1}{2} t\ddot{q}.$$

These equations have, therefore, the following first integral:

$$\Gamma = \sum_{x,y} [(\frac{3}{2} \dot{x} + \frac{1}{2} t\ddot{x}) \delta x - (x + \frac{1}{2} t\dot{x})\delta\dot{x}].$$

We computed the one-arc symmetric periodic solutions of
SDS3. They again have initial conditions of the form $(x_o,0,0,\dot{y}_o)$
The present system appears very different from the two previous
systems SDS1 and SDS2, especially in the neighborhood of the origin, because the potential of SDS3 is homogeneous and has no
second-degree terms. We only have cubic terms here. This system
can thus not be linearized in the ordinary sense in the vicinity
of its only equilibrium point (0,0).

The principal conclusion of the numerical calculations,
which all correspond to $\alpha = 1$, is that there exist <u>four</u> important
families of simple periodic solutions, all originating at the
origin (0,0). We have designated these families by the letters
A,B,C,D, and we will briefly describe each one of them below.

Figure 9 shows the relative positions of the four families in the $(x_o, \dot{y}_o)$-diagram.

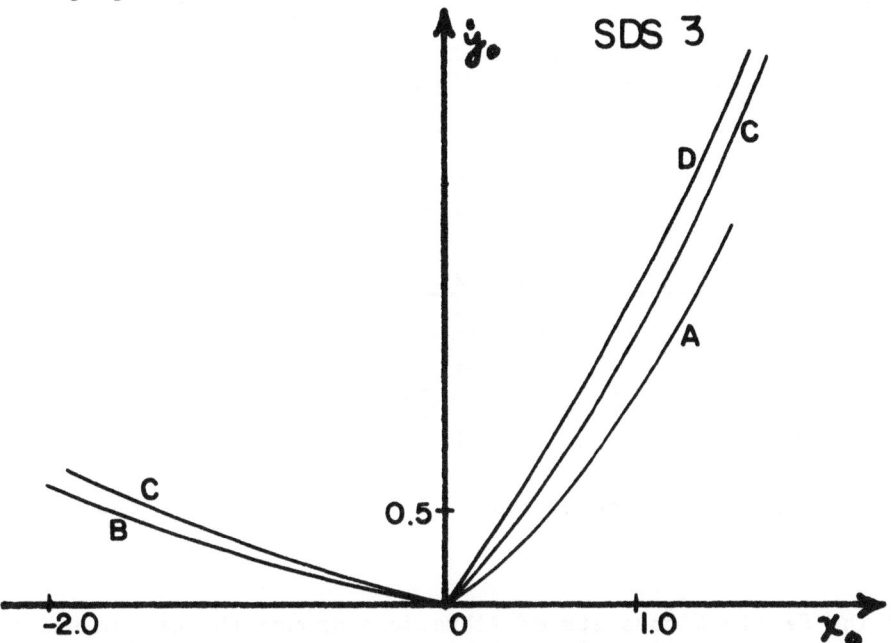

Figure 9. System 3; Initial conditions of periodic solutions.

The family A consists of more or less vertical oscillations originating from the origin as is seen in Figure 10. Note that the orbits intersect the x-axis (at a right angle) only at a single point. Also note that the solutions have a zero-velocity point at both end points.

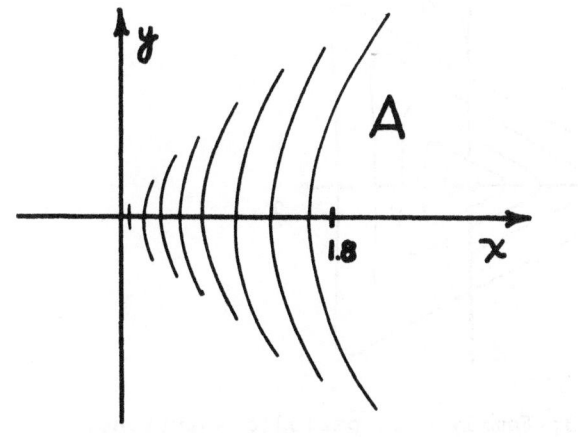

Figure 10.

System 3; Family A of periodic solutions.

The family B also consists of oscillations originating at the origin but here the solutions intersect the negative x-axis (Figure 11).

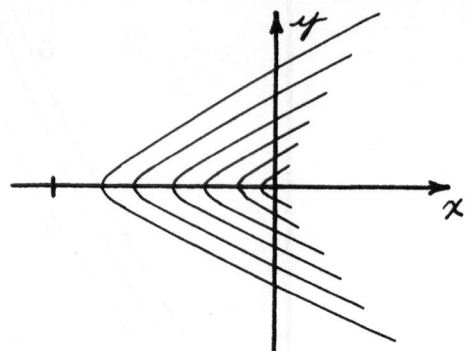

Figure 11. System 3; Family B of periodic solutions.

The family C consists of librations around the origin which are more or less triangular in shape. Again, this family begins at the origin (Figure 12).

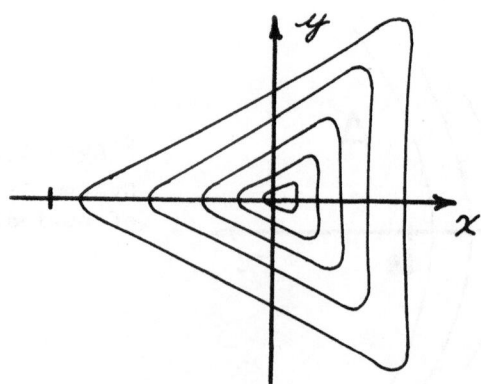

Figure 12. System 3; Family C of periodic solutions.

The family D consists of oscillations around the origin with
increasing amplitude. The intersections with the x-axis are in
the positive values while the zero-velocity points occur at the
left side of the origin. It should be noted that the initial
velocities of the families C and D differ very little. The same
is true for the families C and B in the negative values of x_o
(Figure 13).

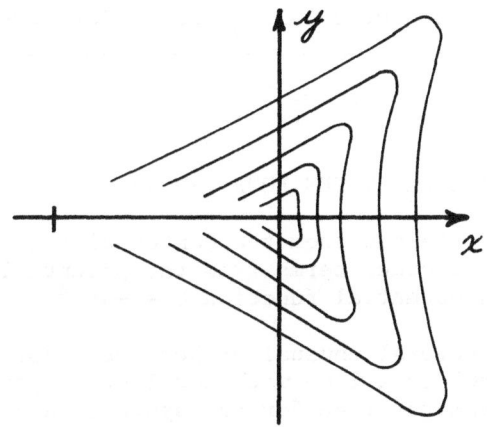

Figure 13. System 3; Family D of periodic solutions.

The most important property of the periodic orbits of this
system is that, within each family, all the orbits are similar to
each other because the potential of this problem is homogeneous.
The solutions of a family differ only by a scale factor. It is
thus sufficient to know the initial conditions of a single solution in order to be able to obtain those of all the others of the
same family. More precisely, if we multiply the abscissa x_o
by λ, we must multiply the initial velocity $\dot{y}_o$ by $\beta = \lambda^{3/2}$ and
the period will then be changed by a factor $\beta = \lambda^{-1/2}$. In light
of this remark, it will thus be sufficient to give the initial
conditions of one solution of each family. The following table
gives five quantities for one solution of each family. T represents the period, E the energy, and k the stability index.

x_o	$\dot{y}_o$	T/2	E	k	family
0.2	0.21053980	4.6067543	0.01949684	752.10	A
0.2	0.24853083	13.176049	0.02821712	3.9E+8	D
0.13930056	-0.14357502	7.9607004	0.00940587	-19485	C
-0.3	0.02849104	7.9607004	0.00940587	-19485	C
-0.3	0.03068610	10.504653	0.00947082	538000	B

6.   THE SIMPLE DYNAMICAL SYSTEM FOUR (SDS4)

This system is obtained from the Contopoulos system (SDS2) by removing the second-order terms from the potential. This is thus the system with potential function $U = -\alpha xy^2$.

This sytem has several unusual properties. For instance, the whole x-axis (y=0) is a line with equilibrium points. The x-axis is again a symmetry line for the system, as can be seen by comparison with the general form of symmetric systems.

Let us show an important property of the present system. We will prove that the system has no periodic solutions but that all solutions escape; when t goes to infinity, so do x and y. We exclude here the equilibrium points y=0. We first prove that in all cases x becomes negative as t increases. We assume the parameter α positive. The first equation of motion shows that $\ddot{x}$ is always < 0. Thus $\dot{x}$ is always decreasing and finally becomes < 0. In other words, for all initial conditions such that $\alpha > 0$, y≠0 for large enough values of t, x will become negative.

Let us now show that a consequence of this fact is that all solutions escape. For this purpose, we use the quantity

$$J = \frac{1}{2} r^2 = \frac{1}{2} (x^2 + y^2),$$

and we will show that for all solutions $\ddot{J}$ becomes, and stays, positive. We have, indeed,

$$\dot{J} = r\dot{r} = x\dot{x} + y\dot{y},$$

$$\ddot{J} = r\ddot{r} + r^2 = x\ddot{x} + y\ddot{y} + \dot{x}^2 + \dot{y}^2.$$

To simplify this result, we use the equations of motion and the energy equation

$$\dot{x}^2 + \dot{y}^2 = 2(E - \alpha xy^2).$$

we find, then

$$\ddot{J} = 2E - 5\alpha xy^2 = 2E + 5U.$$

What we have here is known as the Lagrange–Jacobi identity and is in fact a consequence of the property that the potential U of the problem is homogeneous of degree 3 in the coordinates. We know that another form of the identity is

$$\ddot{J} = 2T + 3U = 5T - 3E.$$

the most interesting form is probably

$$\ddot{J} = \dot{x}^2 + \dot{y}^2 - 3\alpha xy^2.$$

This shows that for $\alpha x < 0, \ddot{J}$ is positive. Consequently, J increases and finally becomes positive. Then J increases forever. This does not mean that the function J (or r^2) can have only one minimum. The function J may oscillate a few times before going to ∞. It is because of this property that we have done very little numerical calculations on this dynamical system; all solutions are escape solutions and it has no periodic solutions. We include a single figure which illustrates a typical solution, with initial conditions (-0.8, 0) and initial velocity (0, 1.05)(Figure 14).

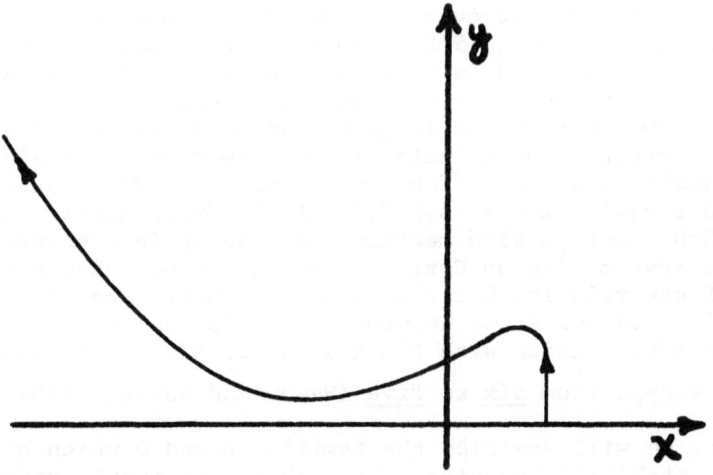

Figure 14. System 4; An escape trajectory in SDS4.

7.   THE STABILITY OF THE SOLUTIONS OF THE HÉNON SYSTEM.

We give here a more detailed account of the stability of the simple periodic solutions of the five families A,B,C,D,E   which have been described in section three.   The families A and E will be disposed of first because of their simplicity, all the orbits being unstable.

At the origin (o,o) of the family A, the stability indes k (the sum of the two non-trivial eigenvalues λ and $1/\lambda$ of the monodromy matrix) is equal to +2 and from there on it increases steadily.   For instance at $x_0 = -1.9$ k has reached the value 7285 and these periodic solutions are thus very unstable.   Remember that stability corresponds to the values −2 to +2 of k (Whittaker, 1960, page 404).

Similarly, the family E is also unstable.   At the origin of the family, at $x_o = 1.0$, the stability index k starts at 38.8 and increases steadily from there on.   At $x_o = 1.6814$, $\dot{y}_o = 3.4232$, we have the value k = 129.   There are no critical orbits and no branchings in families A and E, because there are no stable orbits.

The family C of periodic solutions has a somewhat more interesting evolution.   At the origin (o,o), it begins with a stability index k equal to +2.   The stability curve is tangent to the horizontal line k = +2 at the origin and from there on it decreases continuously towards the negative unstable values.   The transition k = −2 is reached at the critical periodic solution (Henon, 1965) with initial conditions $x_o = -0.40575$, $\dot{y}_o = 0.2967$. The monodromy matrix of this critical orbit has two eigenvalues equal to +1 and two others equal to −1.   It is thus definitely of the second kind (Henon, 1965).   In fact, the analysis of the four coefficients a,b,c,d of the Henon matrix shows that we are in presence of a solution of type six, because we find a=d=−1 and b=0.   Also there is no extremum of the Jacobi Constant at this orbit.   These facts are comfirmed by the half-revolution analysis with the Henon-Guyot method.   We found that among the four coefficients A,B,C,D of the Guyot-matrix, we have here D=0.   Let us also mention that, being in presence of a reversible system with no Coriolis forces, we may change the direction of the velocity but this does not change the type (six) of the orbit.   However if we change the initial point of the orbit to the other intersection with the x-axis (x_1 =0.3252, $\dot{y}_1$ =.4631), we change the type from six to five (Henon and Guyot, 1970).

Finally, we will describe the families B and D which have a very similar stability behavior.   The orbits are mostly unstable, except in a very narrow region in the center of the family.   The

family B is unstable (with k > 2) outside the interval
x_o =-0.0972 to -0.0775 and stable inside. At these two limits
we have critical orbits of the first kind. The first one, at
x_o =-0.0972, $\dot{y}_o$ = .70467, is of type three (c = 0, C = 0) and
is the bifurcation with the family D. The other limit is the
critical orbit, at x_o =-0.0775, $\dot{y}_o$ = 0.70675, which is either
of type one or of type three, with c=C=0 and an extremum of
the Jacobi Constant.

Another remarkable critical orbit of the family B was found
at x_o = -.0875, $\dot{y}_o$ = .70563, where k is minimum and tangent to
the line k = -2. This seems to be a critical orbit of the combined types five and six, because we have A=D=b=c=0, (within
the precision of the numerical integration!).

The family D, which in fact branches off the family B, displays a quite analogous behavior. It is unstable (with k < -2)
outside the interval x_o = -.0.1060 to -0.0882 and stable inside. The critical orbit is the same orbit at both ends. It
appears of type six at x_o = -.1060 and of type five at
x_1 = -.0882 (with k = -2).

In the middle of the stable part of family D, k reaches the
maximum +2 (tangent to k = +2) at the intersection with the
family B. We have a minimum of the Jacobi constant at this
periodic orbit and the type is three. At the same point the
period goes through a maximum.

We have only described here the most simple periodic solutions. It is clear that more complicated solutions can be found,
with multiple loops, by starting from the stable orbits of the
present families. All the stability calculations have been performed with three different methods: the integration of the variational equations, the Henon method (1965) and also the Henon-
Guyot method (1970).

8.   REFERENCES.

Cherry, T.M.(1924), Proc. Camb. Phil. Soc., Vol. 22, pp. 287,
       325, 510.

Contopoulos, G. (1965), Astron. J. Vol. 70, No. 8, pp. 526-544.

Hénon, M. (1965), "Numerical Explorations of the Restricted Pro-
       blem", (Part 2), Annales d'Astrophysique, Vol. 28, No. 6,
       pp. 992-1007.

Hénon, M. and Guyot, M. (1970), "Stability of Periodic Orbits
    in the Restricted Problem", in Periodic Orbits, Stability
    and Resonances, G. Giacaglia (ed.), Reidel Publ. Co.,
    pp. 349-374.

Hénon, M. and Heiles, C.E. (1964), Astron. J., Vol. 69, No. 1,
    pp. 73-79.

Losco, L. (1977), Cel. Mech. 15, pp. 477-488.

Poincaré, H. (1892), Methods Nouvelles 1, Chapter 5.

Whittaker, E.T. (1960), "Analytical Dynamics", Cambridge Univ.
    Press.

Wintner, A. (1947), "Celestial Mechanics", Princeton Univ. Press.

INSTABILITIES IN SYSTEMS OF THREE DEGREES OF FREEDOM

G. Contopoulos

University of Athens, Greece.

INTRODUCTION.

Dynamical Problems of three-degrees of freedom are important for two main reasons:

(1)   They are the simplest problems where qualitatively new phenomena, beyond those of two-degrees of freedom, are expected. For instance, according to the KAM theorem, a linearly stable periodic orbit is, in general, stable in two dimensional systems, while it is expected to be unstable in systems of three or more dimensions.

(2)   There are several three-dimensional problems of actual interest, like the general motion of a star in a galaxy, the plane general three-body problem, the three-dimensional restricted three-body problem, etc.

For these reasons theoretical and experimental studies of systems of three-degrees of freedom are of great interest.   Such studies are expected to give general understanding of systems of three or more degrees of freedom.

In this context we are going to discuss:

(1)   The similarities and differences between systems of two and three degrees of freedom, and (2) how a systematic study of a given three-dimensional system may be made.

Victor G. Szebehely (ed.), Instabilities in Dynamical Systems. 25-39.

1.   SYSTEMS OF (2+1)-DEGREES OF FREEDOM.

Systems of two-degrees of freedom are well understood today. One starts with an integrable system and introduces a perturbation and the system becomes non-integrable. For small perturbations the non-integrable system is very close to the integrable one. However, new phenomena appear, the most important being the intersection of the asymptotic surfaces emanating from the unstable periodic orbits along homoclinic and heteroclinic orbits, and the subsequent appearance of small "zones of instability" (Birkhoff, 1927).

When the perturbation becomes large these "zones of instability" merge to produce large ergodic, or stochastic "seas", which may, eventually, cover the whole available space.

The simplest systems in three-dimensions consist of a two-dimensional motion plus an independent motion along a third dimension. Such systems are partially separable. It is obvious that such systems do not present anything essentially new, beyond what refers to the two-dimensional motion.

The separation of variables may be realized only after a change of variables. Such is the case studied by Ford (1975):

$$H=I_1+2I_2+3I_3+\varepsilon[\kappa I_1 I_2^{1/2}\cos(2\vartheta_1-\vartheta_2)+ (I_1 I_2 I_3)^{1/2}\cos(\vartheta_1+\vartheta_2-\vartheta_3)] \ ,$$

$$(1)$$

where (I_i,ϑ_i) are action-angle variables and $\kappa,\lambda,\varepsilon$ are constants. Ford remarks that the combination $I_1+2I_2+3I_3$ is an exact second integral of motion, but that there is no third integral in general (for arbitrary κ and λ). Thus the motion is stochastic, in general, and this is true however small the perturbation ε.

In fact, if we perform a canonical change of variables by means of the generating function

$$S=J_1(2\vartheta_1-\vartheta_2)+J_2(\vartheta_1+\vartheta_2-\vartheta_3)+J_3\vartheta_3 \ ,$$

$$(2)$$

we find new action-angle variables (J_i,ψ_i) by the relations:

$$I_1=2J_1+J_2 \ , \qquad I_2=J_2-J_1 \ , \qquad I_3=J_3-J_2 \ ,$$

$$(3)$$

$$\psi_1 = 2\vartheta_1 - \vartheta_2 \ , \quad \psi_2 = \vartheta_1 + \vartheta_2 - \vartheta_3 \ , \quad \psi_3 = \vartheta_3 \ , \tag{4}$$

and the Hamiltonian (1) becomes

$$\tag{5}$$

$$H = 3J_3 + \varepsilon \{ \kappa (2J_1 + J_2)(J_2 - J_1)^{1/2} \cos\psi_1 + \lambda [(2J_1 + J_2)(J_2 - J_1)(J_3 - J_2)]^{1/2} \cos\psi_2 \}.$$

This Hamiltonian does not contain the angle ψ_3, therefore, J_3 is an integral of the motion. Thus the motion of the two-dimensional system, $(J_1, J_2, \psi_1, \psi_2)$ is independent of the motion along the third degree of freedom (J_3, ψ_3). If the two-dimensional motion is stochastic, it remains so if $\varepsilon \neq 0$ varies. A change of ε does not change the orbits, only the time scale that is needed to describe the orbits. Therefore the system (1) is completely integrable if $\varepsilon = 0$ while it is stochastic for arbitrary small ε.

This case is considered by Ford as a new type of integrability. However we notice that this behaviour is directly related to the partial separability of the system (1), or (5). That means that the system (1) is a (2+1)-dimensional system, i.e., it can be reduced to a two-dimensional motion, plus an oscillation along the third dimension.

A similar behaviour occurs in any Hamiltonian, which contains only two combinations of angles:

$$H = H[I_1, I_2, I_3, \cos(m_1\vartheta_1 + m_2\vartheta_2 + m_3\vartheta_3), \cos(n_1\vartheta_1 + n_2\vartheta_2 + n_3\vartheta_3)] \ ,$$

$$\tag{6}$$

where m_i, n_i are integers not all zero, and such that the m_i are not all proportional to the corresponding n_i.

The system (1) of Ford has one extra characteristic, namely if we write its linear part in the form $(\omega_1 I_1 + \omega_2 I_2 + \omega_3 I_3)$, we have

$$\omega_1' = \omega_1 m_1 + \omega_2 m_2 + \omega_3 m_3 = 0 \ . \tag{7}$$

$$\omega_2' = \omega_1 n_1 + \omega_2 n_2 + \omega_3 n_3 = 0 \ . \tag{8}$$

In our case $\omega_1=1$, $\omega_2=2$, $\omega_3=3$, $m_1=2$, $m_2=-1$, $m_3=0$, $n_1=n_2=1$, $n_3=-1$.

The problem arises as to what happens in a system which is near to, but not exactly at such a double resonance.

If ω_1', ω_2' are not zero, then the Hamiltonian (5) becomes:

$$H=\omega_1'J_1+\omega_2'J_2+3J_3+\varepsilon F(J_1,J_2,J_3,\cos\psi_1,\cos\psi_2) \ . \qquad (9)$$

The action J_3 is always an integral of the motion, but in this case there is one more asymptotic integral of the motion if ε is sufficiently small, e.g., if ε/ω_1' is smaller than a limiting value ε_d there is only a small degree of stochasticity in the system, while if $\varepsilon/\omega_1' > \varepsilon_d$ the system is almost completely ergodic. The transition from small to large stochasticity is abrupt. This was found, among others, by Henon and Heiles (1964), and by Barbanis (1966).

If $\omega_1'=0$ but $\omega_2'\neq0$ and $\dfrac{\varepsilon}{\omega_2'} < \varepsilon_d$ again there is no appreciable stochasticity. This means that a single resonance does not introduce stochasticity. Only in the case of a double resonance $\omega_1'=\omega_2'-0$ can stochasticity appear for any .

Thus we can state that stochasticity sets in if the perturbation is large enough with respect to the resonant combinations of frequencies, ω_1' and ω_2'.

2.   INVARIANT SURFACES.

Let us consider now a general three-dimensional Hamiltonian

$$H = H(I_i,\vartheta_i) = h \ , \qquad (10)$$

which is "near" an integrable one.

Then, according to the KAM-theorem, we have a set of integral surfaces, given by equation (10) and

$$\Phi_1 = \Phi_1(I,\vartheta) = c_1 \ , \qquad (11)$$

$$\Phi_2 = \Phi_2(I_i,\vartheta_i) = c_2 \ . \qquad (12)$$

Eliminating I_2 and I_3 from these equations we find

$$f(I_1, \dot{v}_1, \dot{v}_2, \dot{v}_3) = 0 , \qquad (13)$$

where f contains also h, c_1 and c_2. This is an "integral surface" in the four-dimensional space $(I_1, \dot{v}_1, \dot{v}_2, \dot{v}_3)$.

One way to represent this surface is by taking sections $\dot{v}_3 = $ const. In particular we may use a section $\dot{v}_3 = 0$ and find an "invariant surface"

$$f(I_1, \dot{v}_1, \dot{v}_2, 0) = 0 , \qquad (14)$$

which correspond to an "invariant curve" of a two-dimensional problem.

The surface (13) contains the orbits whose initial conditions satisfy the relations (10), (11) and (12); therefore, the invariant surface (14) contains the points of intersection of all these orbits with the plane, $\dot{v}_3 = 0$.

However, in the non-integrable cases the surfaces (13) and (14) do not form a continuum. This means that there are orbits that do not lie on integral surfaces of the form (13). Such orbits may cross a given surface (13), if at the points of intersection the values of I_2, I_3 do not satisfy the relations (11), (12) with the given values of c_1, c_2. Similarly, the points $(I_1, \dot{v}_1, \dot{v}_2)$ with $\dot{v}_3 = 0$ may lie on both sides of a given surface (14).

The existence of surfaces of the form (14) can be found empirically if we have many points $(J_1, \dot{v}_1, \dot{v}_2)$ and check whether they are on a simple surface or not. Froeschlé (1970) in the case of the three-dimensional restricted three-body problem gave examples of existence and non-existence of invariant surfaces of this type, by using stereoscopic views of the distribution of the points on a three-dimensional "surface of section". More examples of this type would be very useful.

Certain cases where the equation of the surfaces (14) can be given explicitly, will be discussed first. Then more general cases will be treated.

<u>Case 1</u>.   The Hamiltonian is a function of the actions only:

$$H=H(I_1,I_2,I_3) \ . \tag{15}$$

In this case the invariant surface is the plane

$$I_1 = \text{const. ,} \tag{16}$$

while the angles ϑ_i vary linearly in time

$$\vartheta_i = \frac{\partial H}{\partial I_i} \, t + \vartheta_{i0}. \tag{17}$$

The successive points on the invariant surface (called "consequents" by Poincaré) are given by the relations

$$\vartheta_i = 2K\pi \left(\frac{\partial H}{\partial I_i} \right) \Big/ \left(\frac{\partial H}{\partial I_3} \right) + \vartheta_{i0} \ (\text{mod} 2\pi) \ , \quad (i=1,2) \ , \tag{18}$$

where   K   is an integer $0,1,2,\ldots$.

Depending now on whether the ratios

$$R_i = \frac{\partial H}{\partial I_i} \ \Big/ \ \frac{\partial H}{\partial H_k} \quad (i,j,k=1,2,3; \ 2,3,1; \ 3,1,2) \tag{19}$$

are rational or irrational, we distinguish the following cases.

<u>Case 1a</u>. R_i=irrational   (i=1,2,3) .

This is the general case.   The consequents fill the plane I_1=constant densely.

<u>Case 1b</u>.   One R_i is rational, the other irrational.

In this case, the consequents are dense along a certain number of lines.   By changing ϑ_{i0}, we have similar sets of lines. The lines, corresponding to all the values of ϑ_{i0} from 0 to 2π, fill the surface (16).

<u>Case 1c</u>.   All R_i are rational.

In This case, there are only a finite number of points of intersection on the plane I_1=constant that represent periodic orbits.

By changing w_{10} we have similar sets of points. Therefore, the periodic orbits are dense on the plane $I_1 =$ const.

Case 2. In this case we have, in the Hamiltonian, one (but only one) combination of angles, besides the actions, e.g., we may have

$$[H = H_1 \; I_1, I_2, I_3, \; \cos(m_1 w_1 + m_2 w_2 + m_3 w_3)] \; . \tag{20}$$

This case is also integrable. A well known canonical change of variables by means of the generating function

$$S = J_1(m_1 w_1 + m_2 w_2 + m_3 w_3) + J_2 w_2 + J_3 w_3 \; , \tag{21}$$

gives

$$I_1 = m_1 J_1, \quad I_2 = m_2 J_1 + J_2, \quad I_3 = m_3 J_1 + J_3 \; , \tag{22}$$

$$\psi_1 = m_1 w_1 + m_2 w_2 + m_3 w_3, \quad \psi_2 = w_2, \quad \psi_3 = w_3 \; , \tag{23}$$

and

$$H = H_1(J_1, J_2, J_3, \cos\psi_1) \; .$$

Since the angles ψ_2, ψ_3 are ignorable the actions J_2, J_3 are integrals of the motion; therefore, the Hamiltonian is reduced to a one-dimensional case. Equation (24) represents orbits on the plane (J_1, ψ_1). Alternatively, if we consider the two-dimensional problem $(J_1, J_3, \psi_1, \psi_3)$ and mark the points (J_1, ψ_1) when $\psi_3 = 2K\pi$, the same curves (24) are invariant curves.

In general the invariant curves on the plane (J_1, ψ_1) are of two types: (i) curves going all the way from $\psi_1 = -\pi$ to $\psi_1 = +\pi$ and (ii) closed islands, when ψ_1 varies from a minimum to a maximum value (see such an example in Fig. 1). A limiting case is the set of invariant points that represent stable periodic orbits (like S_1). The two types of invariant curves are separated by the closed asymptotic curves going through the unstable invariant points like U_1, U_2.

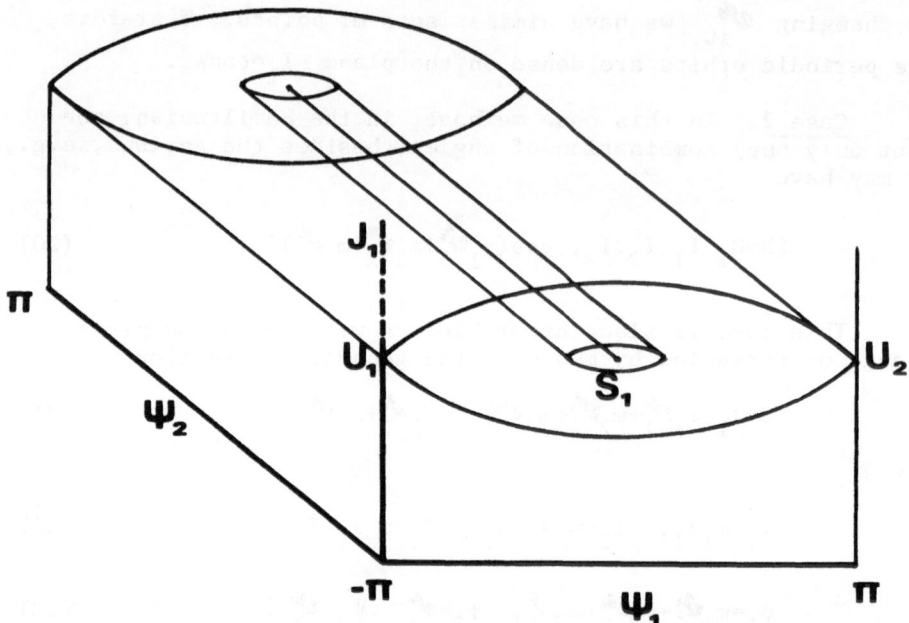

Fig. 1.　Invariant surfaces in an integrable Hamiltonian, containing only one angle (Case 2). The periodic orbits are represented by lines through the points S_1 (stable), or $U_1 U_2$ (unstable).

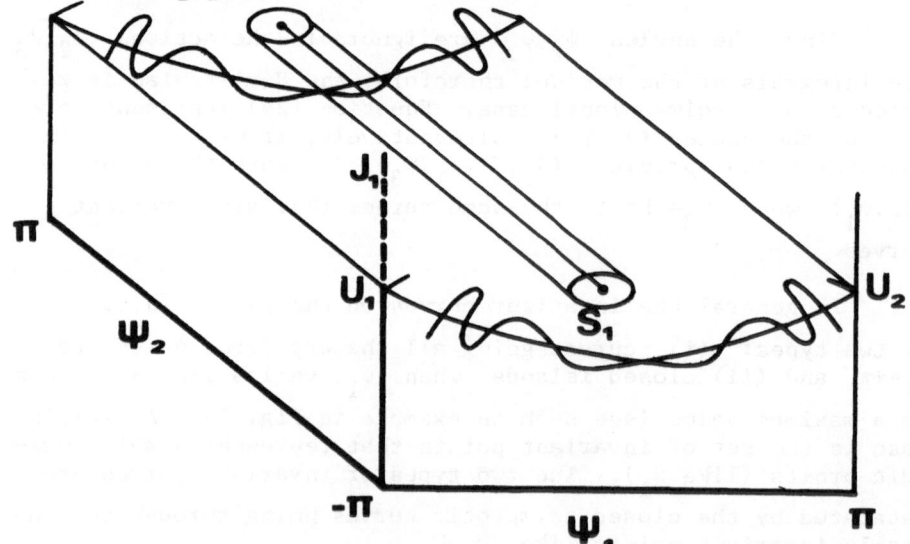

Fig. 2.　Invariant surfaces in a particularly separable Hamiltonian, containing two angles (Case 3). The periodic orbits are as in Fig. 1.

The corresponding invariant surfaces of the three-dimension problem are cylindrical, with ψ_2 varying from $-\pi$ to $+\pi$.

We distinguish again special cases similar to those of case 1.

The successive consequents for $\psi_3 = 2K\pi$, in general, give points densely distributed on a cylinder (case 2a). This is due to the fact that the derivatives of ψ_1,

$$\frac{d\psi_1}{dt} = \frac{\partial H_1}{\partial J_1} \qquad (25)$$

are not constant, in general, but depend on the value of ψ_1.

However, ψ_1 may take only a finite number of values. This may happen because on each curve in the (J_1, ψ_1) plane a "rotation number" can be defined, which varies continuously with the distance of the curve from the invariant points S_1; thus, the rotation number goes through rational values. In such a case the consequents lie on a number of lines parallel to the ψ_2-axis (Case 2b). We have a similar case if the consequents lie on a finite number of simple lines on the cylinder.

Finally, if both ψ_1 and ψ_2 take only a finite number of values we have periodic orbits (Case 2c).

Case 3. If the Hamiltonian contains two different linear combinations of the angles $\mathcal{V}_i$, it is of the form (6). Then a canonical change of variables by means of the generating function

$$S = J_1(m_1\mathcal{V}_1 + m_2\mathcal{V}_2 + m_3\mathcal{V}_3) + J_2\mathcal{V}_2 + J_3(n_1\mathcal{V}_1 + n_2\mathcal{V}_2 + n_3\mathcal{V}_3) \qquad (26)$$

gives a Hamiltonian that depends on $\psi_1 = m_1\mathcal{V}_1 + m_2\mathcal{V}_2 + m_3\mathcal{V}_3$ and on $\psi_3 = n_1\mathcal{V}_1 + n_2\mathcal{V}_2 + n_3\mathcal{V}_3$, but not on $\psi_2 = \mathcal{V}_2$.

Thus J_2 is an integral of motion and ψ_2 varies from $-\pi$ to $+\pi$. Therefore, the integral surfaces are cylindrical along the ψ_2-axis (Fig. 2) but in general the problem in the $(J_1\psi_1)$ plane is non-integrable.

In particular, if we have an unstable invariant point on the (J_1, ψ_1) plane, the asymptotic curves intersect along an infinity of homoclinic points. If we take a segment on the outgoing asymptotic curve between·two successive consequents, near the point U_1, we can construct the asymptotic curve by taking the consequents of this segment. Thus we follow the oscillations of the asymptotic curve on the plane (J_1, ψ_1) as we approach the point U_2.

Now consider also the variation of ψ_2. A point close to U_1 on the outgoing cylindrical surface goes through the axis ψ_2 (from the point U_1) and has its consequent on the same cylindrical surface. Thus any arc on this surface joining the two points has as successive consequents arcs that form a curve on the surface approaching asymptotically the axis from U_2 (parallel to ψ_2).

If we have no a-priori knowledge of the asymptotic surface we can construct the surface by taking a small cylindrical area (with $-\pi < \psi_2 < \pi$ on the outgoing asymptotic surface) and by finding its successive consequents.

Case 4. The previous situation is a partially separable (2+1)-dimensional case, with ψ_2 along the separated dimension. We get a closer approach to the general non-separable case if we consider again a (2+1)-dimensional case but in a subspace that does not contain the separated dimension. Namely, we just interchange ψ_2 and ψ_3 and consider a non-integrable two-dimensional system $(J_1, J_2, \psi_1, \psi_2)$ plus an oscillation along ψ_3. In this case the invariant surface $(J_1, J_2, \psi_1, \psi_2)$ is an integral surface on the subspace (J_1, ψ_1, ψ_2) of the phase space, while the conditions $\psi_3 = 2K\pi$ just regulate the distribution of the consequents on this surface.

The integral surfaces in a near integrable system are like distorted double cylinders. Namely, we have an inner and an outer distorted cylinder. The distortion is small in non-resonant cases, while it may be very large in resonant or near resonant cases. Examples of both non-resonant and resonant integral surfaces were given by Barbanis (1966a). It would be useful to extend this work in order to have a more general idea of the form of the integral surfaces, especially in various resonant cases.

In the two-dimensional cases we know that the various resonances change the topology of the invariant curves, so that instead of curves closing around the central invariant point we have islands around the invariant points representing resonant periodic orbits. Similarly, in the three-dimensional cases we have a corresponding change of the topology and the distorted cylinders. What is not well known is the degree of distortion of the various cylindrical surfaces.

Of special interest is the study of the asymptotic invariant surfaces. Such a study would give an explanation of the observed dissolution of the invariant surfaces in these cases.

Case 5. The most general case is when three or more combinations of angles appear at the same time in the Hamiltonian.

We expect that the topology of the integral surfaces in this case is very similar to case 4. However, there is more interaction among the various degrees of freedom than in a (2+1)-dimensional system, e.g., if we have two near integrable Hamiltonians $H_1 = H_1(J_1, J_3, \vartheta_1, \vartheta_3)$ and $H_2 = H_2(J_2, J_3, \vartheta_2, \vartheta_3)$ we have a small degree of dissolution of the invariant curves on the planes (J_1, ϑ_1) and (J_2, ϑ_2) respectively. Consider a three-dimensional Hamiltonian,

$$H = H_1 + H_2 , \tag{27}$$

coupling the two Hamiltonians through their dependence on ϑ_3. We have at the same time a coupling of the dissolution, which may produce a larger three-dimensional dissolution than in the separate planes (J_1, ϑ_1) and (J_2, ϑ_2). This subject is of great theoretical and practical interest, but no systematic study exists up to now. Mr. P. Magnenat (Geneva) has started an exploration of the neighbourhood of unstable periodic orbits in a three-dimensional problem in order to find the degree of dissolution, which is expected to be maximum there.

On the other hand it seems that the appearance of distorted cylindrical surfaces is quite common in rather general three-dimensional dynamical problems. Such are the plane general three-body problem (Hadjidemetriou, 1975; Contopoulos, 1978a), the three-dimensional restricted three-body problem (Froeschlé, 1970), and the problem of three coupled oscillators (Martinet, 1978).

3.   DISJOINT ERGODIC SEAS.

In a non-integrable system there are, in general, regions that contain integral surfaces and regions of stochasticity. From numerical and theoretical experience with two-dimensional systems (see, e.g., Henon and Heiles, 1964; Contopoulos, 1967) we know that for small perturbations the regions of stochasticity are small. However, between any two integral surfaces we have stable periodic orbits, surrounded by sets of higher order islands, and unstable periodic orbits with asymptotic surfaces intersecting along homoclinic points. Therefore, between any two integral surfaces there are regions of stochasticity. In general these ergodic regions are very small, and can be ignored for practical purposes. The regions where stochasticity is below a certain level we call "ordered regions".

On the other hand, for large perturbations the regions of stochasticity are connected and form large "ergodic seas". Inside such seas only small islands of stability remain.

In intermediate cases we have both ordered and stochastic regions. Usually there are only a few disjoint ergodic seas with clearly defined boundaries. If between two ergodic seas there is at least one exact integral surface the two seas are disjoint and do not communicate. The time averages of various quantities in these ergodic regions are different.

We explore now what happens in systems of three or more degrees of freedom.

If the diffusion mechanism described by Arnold (Arnold and Avez, 1967) is operative, then one would expect that all stochastic regions communicate and that an orbit in a given ergodic sea will eventually go through all other ergodic seas of the system. This may take a long time (the Arnold diffusion time) but in the end it will produce an equality of the time averages of any given quantity for practically all initial conditions. Furthermore, if the available phase space extends to infinity then most orbits will find their way eventually to infinity and escape.

We have already given indications that his expected escape does not occur for times much longer than the estimated Arnold diffusion times (Contopoulos, 1978a).

Some further experiments exploring whether the various ergodic seas communicate, or are actually disjoint, were made recently (Carrota, et al, 1978; Contopoulos, Galgani and Giorgilli, 1978). One can decide empirically whether a given orbit is in an ordered region (i.e., departures from an integral surface are very small), or in an ergodic sea by calculating the maximal Liapunov characteristic (Benettin, et al, 1976). This number is calculated as the maximum of the limit $\lambda_m = \lim \chi(t)$, where $\chi(t) = \frac{1}{t} \log\| \xi(t) \|$, and ξ is the solution of the variational equations.

If the largest Liapunov characteristic number is different from zero the orbit is stochastic. On the other hand if the limit of $\chi(t)$ tends to zero, the orbits are in an ordered region. In practice we require that λ_m should be below a certain small level. But this level cannot be arbitrarily small because of the inaccuracies of the calculations.

In our experiments (Contopoulos, et al, 1978) we studied a system of three coupled oscillators

$$H = \frac{\omega_1}{2} (q_1^2 + p_1^2) + \frac{\omega_2}{2} (q_2^2 + p_2^2) + \frac{\omega_3}{2} (q_3^2 + p_3^2) + q_1^2 (q_2 + q_3).$$

(28)

We found an ordered region and 3 separate ergodic regions, one containing many orbits (large sea) and two containing probably smaller sets of orbits (small seas). The three seas seem to be disjoint because the limiting values of λ_m are clearly different in them (Fig. 3).

However in order to establish the disjointness of the three ergodic seas it would be necessary to extend our calculations for much longer times to check whether there is any tendency of λ_m to reach to the same "final" value.

What we found is really curious. If the accuracy of the calculation was relatively small we found that the value of λ_m of the orbit in one small sea started to change, after a certain time, until it reached the value of λ_m of the large sea. This indicates that the corresponding orbit eventually reached the large sea, therefore, the two seas communicated.

However, when the accuracy of the calculations was improved, the time scale for this transition became longer, exceeding our time limits. In order to give an indication of the extremely long time scales involved, note that the time-scale in Fig. 3 is logarithmic. The values of χ in the ergodic seas are stabilized after $10^2 - 10^3$ time units, while the maximum time of our calculations was 2×10^5 time units.

This result leaves us with two alternatives: Either (a)

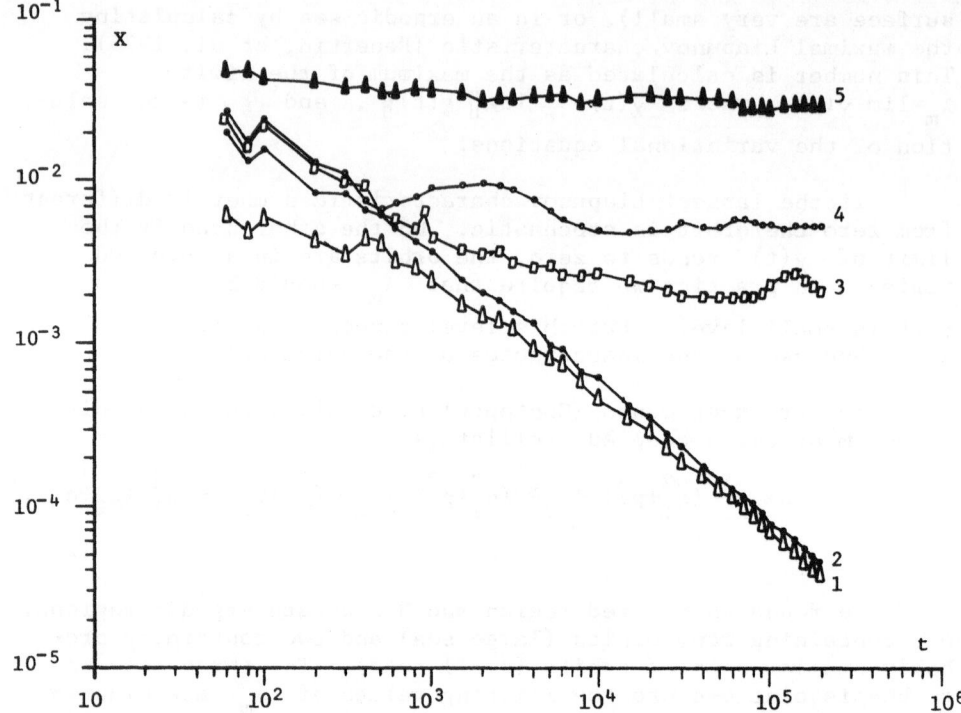

Fig. 3.   The value of $\chi(t)$ as a function of t in the case of 5
          orbits. Orbits 1 and 2 are probably in an "ordered
          region", orbits 3 and 4 in "small seas", and the orbit
          5 in the "large sea".

the ergodic seas do communicate but on a very long time scale,
or (b) they are disjoint but the barrier is in some places thin
enough so that it can be bypassed by the calculation errors.

     One probability is that the ordered regions between the
various ergodic seas do leave some openings that allow communication between the ergodic seas, but these openings have zero
as lower limit, so that the diffusion time is, in fact, infinite.

     In order to visualize what may happen let us consider a
three-dimensional space which contains a nowhere dense set of
one-dimensional integral "surfaces", i.e., lines. For example,
we may consider a room with an infinite number of perpendicular
strings passing through all points with x and y irrational and
"far from rationals" (in the sense of Arnold). Then, in principle, one may go from one end of the room to the other, by following a path close to the rationals. However, the question is how
small is the smallest opening among the strings. If there is a

finite lower bound of these openings, then if one is slender enough he will go through. However, if the lower limit of the openings is zero any finite body will not be able to pass.

It is obvious that further studies of phenomena of this type would be very useful.

REFERENCES.

Arnold, V.I. and Avez, A. (1967), Problèmes Ergodiques de la Mécanique Classique, Gautier Villars, Paris.

Barbanis, B. (1966a), IAU Symp. 25, p. 19.

Barbanis, B. (1966b), Astron. J. 71, p. 415.

Benettin, G., Galgani, L. and Strelcyn, J.M., Phys. Rev. A 14, p. 2338.

Birkhoff, G.D. (1927), Acta Math. 50, p. 359.

Carotta, M.C., Ferrario, C., Lo Vecchio, G. and Galgani, L. (1978), Phys. Rev. A. 17, p. 786.

Contopoulos, G. (1978a), in N. Lebovita, A. Reidel and P. O. Vandervoort (eds.), "Theoretical Principles in Astrophysics and Relativity", Chicago Univ. Press, p. 93.

Contopoulos, G. (1978b), Celes. Mech. 17, p. 167.

Contopoulos, G. (1967), Bull. Astron. 3e Ser. 2, Fasc. 1, p. 223.

Contopoulos, G., Galgani, L. and Georgilli, A. (1978), Phys. Rev. A., (in press).

Ford, J. (1975), in E.D.G. Cohen (ed.), "Fundamental Problems in Statistical Mechanics", Vol. 3, p. 215, North-Holland, Amsterdam.

Froeschlé, C. (1970), Astron. Astrophys. 4, p. 115.

Hadjidemetriou, J. (1975), Celes. Mech. 12, p. 155.

Hénon, M. (1973), Astron. Astrophys. 28, p. 415.

Hénon, M. and Heiles, C. (1964), Astron. J. 69, p. 73.

Martinet, L. (1978), in print.

PERTURBED TWIST MAPS, HOMOCLINIC POINTS AND ERGODIC
ZONES

Robert W. Easton

University of Colorado

ABSTRACT.

   This work studies both numerically and theoretically a two-
parameter family of area preserving mappings of the plane of the
form

$$T : \begin{pmatrix} x \\ y \end{pmatrix} \longrightarrow \begin{pmatrix} a & o \\ o & a^{-1} \end{pmatrix} \begin{pmatrix} \cos(q) & \sin(q) \\ -\sin(q) & \cos(q) \end{pmatrix} \begin{pmatrix} x \\ y \end{pmatrix}$$

where $q = (x^2 + y^2)^b$ and a and b are the parameters. Transformations of this type can occur as Poincaré maps near periodic
orbits of Hamiltonian systems of two degrees of freedom on fixed
energy surfaces. Mappings of this type for $a = 1$ and $b \neq 0$
are called twist mappings. The one-parameter family of mappings
with $a = 1 + \varepsilon$ and b fixed can be considered as the simplest
family of perturbations of the twist map with $a = 1$. For ε
small the K.A.M. theorem implies that T has a collection of
nearly circular invariant curves surrounding the origin. However
for $1.1 \leq a \leq 1.5$ with $b = 1.5$ the origin is a hyperbolic
fixed point whose stable and unstable manifolds intersect trans-
versally, producing homoclinic points. The aim of this work is
to connect the existence of transverse homoclinic points with
the existence of "ergodic zones" (i.e., zones which have positive
area and which are the closure of an orbit). The family of
mappings T(a,b) is particularly simple and exhibits the typical qualitative behavior associated with Hamiltonian systems and
their Poincaré maps.

Victor G. Szebehely (ed.), Instabilities in Dynamical Systems. 41-47.

1. INTRODUCTION.

We study a two-parameter family of area preserving diffeomorphisms of the plane given by

$$T : \begin{pmatrix} x \\ y \end{pmatrix} \longrightarrow \begin{pmatrix} a & o \\ o & a^{-1} \end{pmatrix} \begin{pmatrix} \cos(q) & \sin(q) \\ -\sin(q) & \cos(q) \end{pmatrix} \begin{pmatrix} x \\ y \end{pmatrix}$$

where $q = (x^2 + y^2)^b$. If $a = 1$ and $b \neq 0$ then T is called
a <u>twist map</u> and T rotates each circle centered at the origin
through an angle which depends on its radius. For $a \neq 1$ we
call T a <u>hyperbolic twist map</u> because in this case it is the
composition of a twist map with a hyperbolic linear map. We
wish to study the orbit structure of T where the orbit of a
point (x_o, y_o) is the set of points $\{T^n (x_o, y_o) : n$ is an in-

teger$\}$. The beautiful work of Kolomogorov, Arnold and Moser
(1974) implies that if a is sufficiently close to 1 then the
orbits of most points (in the sense of measure) lie on closed
curves which are slightly distorted circles and these orbits are
dense inside their associated closed curves.

There are many reasons for studying twist mappings and their
perturbations and it is assumed that the reader is familiar with
at least some of these. Hyperbolic twist maps represent the simplest perturbations of twist mappings and my own interest in
these maps arises because they exhibit interesting homoclinic
phenomena and because they can be conveniently studied using a
computer. M. Hénon (1964) has performed numerical studies of
certain transformations and he discovered that some orbits of
these transformations seem to fill densely smooth closed curves
while others seem to fill densely areas with positive measure.
He called these regions "ergodic zones". It is important to ask
whether these zones actually exist or whether they are produced
by something such as round-off error in the computer. It is also
important to ask what mechanism produces these zones if they
exist, and how the transition occurs between the chaotic behavior
in these zones and the regular behavior along nearby invariant
curves. It is my opinion that these zones do indeed exist and
that they are created either by the formation of transverse homoclinic orbits or by the formation of cycles of transverse heteroclinic orbits. Hyperbolic twist maps form a simple class of
examples with which to test the validity of this opinion.

2. RESULTS.

To study the orbit structure of a hyperbolic twist map T it is natural to first look at its local orbit structure near the origin which is a fixed point of T. Define

$$W^s = \{(x,y): \quad T^n(x,y) \longrightarrow (o,o) \quad \text{as} \quad n \longrightarrow +\infty\},$$

$$W^u = \{(x,y): \quad T^m(x,y) \longrightarrow (o,o) \quad \text{as} \quad n \longrightarrow -\infty\} \text{ and}$$

$H = W^s \cap W^u - \{(o,o)\}$. The sets W^s, W^u and H are called respectively the <u>stable manifold of (o,o)</u>, the <u>unstable manifold of (o,o)</u>, and the set of points <u>homoclinic</u> to (o,o). Near (o,o) the transformation T is very close to the linear map determined by the matrix

$$\begin{pmatrix} a & o \\ o & a^{-1} \end{pmatrix}$$

and consequently standard techniques show that W^s and W^u are immersed lines which are tangent at (o,o) to the y-axis and to the x axis respectively. To see the structure of W^s and W^u one can use the computer as follows.

Take m evenly spaced points in an interval $[-\alpha,\alpha]$ along the x-axis and take m evenly spaced points in an interval $[-\beta,\beta]$ along the y-axis. Apply T^k to the points in $[-\alpha,\alpha]$ and apply T^{-k} to the points in $[-\beta,\beta]$. For $a = 1.1$, $b = 1.5$, $\alpha = \beta = .15$, $m = 5000$, $k = 10$ we see approximately the picture for W^u in Figure 1. Notice that the homoclinic points show clearly.

One can also perform the experiment of starting with an initial point (x_o, y_o) and plotting the points $T^k(x_o, y_o)$ for $k = 1,\ldots,m$. Figure 2 was obtained by choosing $(x_o, y_o) = (.1, 0)$ and $m = 2000$. The results of this experiment are rather similar to those obtained by Hénon (1964). However for $a = 1.4$ and $b = 1.5$, $x_o = .3$, $y_o = .3$, $m = 5000$ we receive an error message indicating that $T^k(x_o, y_o)$ is too large a number to be handled by the computer for some k less than 5000. It should be interesting to find the minimum value of a (with b fixed say at 1.5) for which unbounded orbits occur. It would also be interesting to identify the mechanism if there is one) which causes the transition from bounded to 1.5) for which unbounded orbits occur. It would also be interesting to identify the mechanism (if there is one) which causes

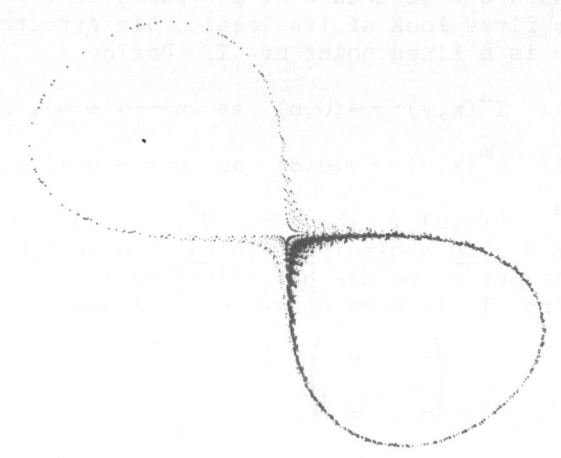

Figure 2

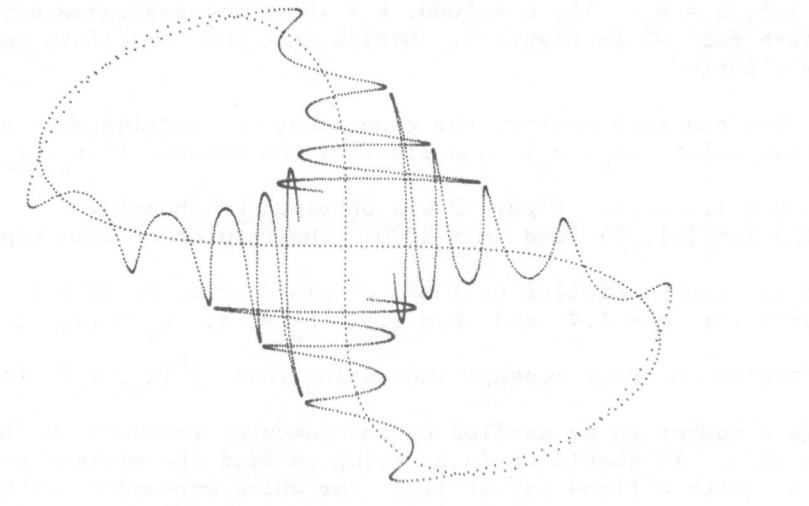

Figure 1

the transition from bounded to unbounded orbits.

Of course numerical experiments can not establish the existence of ergodic zones or the existence of stability and I want
to offer some reasons why homoclinic orbits might be associated
with ergodic zones.

F. Takens (1972) has shown that for a generic, area preserving transformation f of a compact Riemanian manifold M
one has $cl(W^s(p)) = cl(W^u(p)) = cl(H(p))$ where $cl(X)$ denotes
the closure in M of a set X and where p is a hyperbolic
fixed point of f. C. Conley and R. Bowen have defined
ε-chains or pseudo orbits (say for a diffeomorphism g of M)
to be sequences of points $\{x_n\}$ such that $d(x_{n+1}, g(x_n)) < \varepsilon$

for all n. Here $d(x,y)$ denotes the distance in M between
x and y. Pseudo orbits are much like orbits which are obtained
numerically with a computer having a small round-off error.
Conley has shown that as a consequence of Poincaré's recurrence
theorem, given any $\varepsilon > 0$ and points p and q of M there is
an ε-chain $x_0, \ldots, x_N$ with $x_0 = p$ and $x_N = q$. Thus for

area preserving transformations one might expect each numerically
obtained orbit to eventually enter any fixed region. Sometimes
it may be possible to keep track of the total accumulated error
when computing an orbit. This suggests the concept of a strong
ε-chain which is a sequence

$$\{x_k: \ k = 0,\ldots,n\} \ \text{ such that } \sum_{k=1}^{n} d(x_k, g(x_{k-1})) < \varepsilon\} \ .$$

In [3] I have studied strong ε- chains. Suppose that
g has the property that given any points x and y of M and
any $\varepsilon > 0$ there exists a strong ε-chain from x to y.
Then it is easy to show that g has no integrals which are
Lipschitz functions. That is to say if F: $M \longrightarrow R^1$ is such
that $|F(x) - F(y)| \leq L\, d(x,y)$ and $F(g(x)) = F(x)$ for each
xεM then F must be constant everywhere because for arbitrary
points x and y we have a strong ε-chain $\{x_0, \ldots, x_N\}$ with

$x_0 = x$ and $x_N = y$ and consequently

$$|F(x) - F(y)| \leq \sum_{k=1}^{N} |F(x_k) - F(x_{k-1})|$$

$$\leq \sum_{k=1}^{N} |F(x_k) - F(g(x_{k-1}))|$$

$$\leq \sum_{k=1}^{N} L\, d(x_k,\, g(x_{k-1})) \leq L\epsilon \; .$$

Since ϵ is arbitrary it follows that $F(x) = F(y)$. Hence the behavior of orbits is quite complicated in this case.

When $cl(W^s(p)) = cl(W^u(p)) = cl(H(p))$ then one can show that there exists strong ϵ-chains for arbitrary $\epsilon > o$ between any pair of points in $cl(H(p))$ as follows· To ϵ-chain between x and y choose $y' \epsilon W^u(p)$ such that $d(y',y) < \epsilon/4$. Choose N such that $d(T^{-N}(y'), p) < \epsilon/4$. Choose $x' \epsilon W^s(p)$ such that $d(x',x) < \epsilon/4$. Choose M such that $d(T^M(x'),p) < \epsilon/4$. Then

$$\{x, x', T(x'), \ldots, T^M(x'), T^{-N}(y'), \ldots, T^{-1}(y'), y', y\}$$

is a strong ϵ-chain from x to y.

This gives some indication that orbits obtained numerically may go everywhere in the set $cl(W^u(p))$ even when the total error in computing the orbit is arbitrarily small. If it happens that $cl(W^u(p))$ has positive measure, then this set might form an "ergodic zone" where individual orbits fill out regions having positive area or volume. I conjecture that this is the case for the family of hyperbolic twist maps T depending on the parameters a and b for $1.1 \leq a \leq 1.5$ and $b = 1.5$ (and perhaps also for most other values of these parameters).

I also conjecture that T restricted to the set $cl(H)$ is ergodic in the measure theoretic sense. It would be quite interesting to have a precise geometric description of how the transition occurs between the chaotic region $cl(H)$ and the family of smooth invariant curves given by the K.A.M. theorem.

3.   ACKNOWLEDGEMENTS.

Partial support of this work was provided by the National Science Foundation, Grant No. MCS 76-84420.

4.   REFERENCES.

Bowen, R., "On Axiom A Diffeomorphisms", C.B.M.S., Regional
     Conference Series in Mathematics.

Conley, C., "Isolated Invariant Sets and the Morse Index",
     C.B.M.S. Regional Conference Series in Mathematics.

Easton, R. (1978), "Chain Transitivity and the Domain of Influ-
     ence of an Invariant Set", to appear.

Hénon, M. and Heiles, C. (1964), <u>Astronomical Journal</u> <u>69</u>, p. 73.

Moser, J. (1974), "Stable and Random Motions in Dynamical
     Systems", Annals of Math. Studies, No. 77, Princeton Uni-
     versity Press.

Takens, F. (1972), <u>Inventiones Math.</u> <u>18</u>, p. 267.

BIFURCATION THEORY AND ITS APPLICATIONS

Okan Gurel

IBM Corporation, White Plains, New York, USA

SUMMARY. Bifurcation analysis is discussed by summarizing the basic notions, the concept and the elements of bifurcation theory. Applications of bifurcation analysis in various scientific disciplines are listed and briefly outlined.

1. BASIC NOTIONS.

1.1 Systems with Multiple Solutions

The bifurcation phenomenon is a generic concept corresponding to a subset of possible topological changes which might characterize the solution of a system. In general the solution space under consideration, denoted by X, might be endowed with a proper topology. In particular we refer to the euclidean n-space, E_n, a specific metric space. However, in theory, spaces may be considered more general than E_n.

The systems represented by a mapping $f: X \to X$, e.g. $X \subset E_n$, may be analyzed globally by referring to the qualitative behavior of the critical solutions belonging to X. The critical solutions are defined as X_o such that $f: X_o \to X_o$. Taking f as a vector field acting on X, and t as the time element, for example, $x_o = X_o$, $f: x_o \to x_o$, a singular solution of the vector field or $f: x(0) \to (0 + T)$ a periodic solution with a period T, where $x(0), x(0 + T)$ are elements of X_o, and X_o is the

49

Victor G. Szebehely (ed.), Instabilities in Dynamical Systems. 49-60.
Copyright © 1979 by D. Reidel Publishing Company.

critical solution.

By definition, in order to observe the bifurcation
phenomenon, the pair (X,f) must possess varying
multiple solutions.

1.2   Systems Depending on Parameters

Another concept in bifurcation phenomenon is the dependence
on a set of parameters. The mapping $f: X \rightarrow X$ possesses a rather
rigid set of critical solutions. However, if the parameters enter
into this mapping as $f:(X,P) \rightarrow (X)$, where $P \subset E_m$, the analysis

of the solution space based on the variations in P may reveal
various interesting behaviors. The parameter space, P, splits
into two basic parts $P_b \oplus P_{nb}$ where only those parameters

$P_i \in P_b$ play an instrumental role in the analysis of systems

based on the bifurcation theory.

By definition, in order to observe the bifurcation
phenomenon, the triple (X,P,f) must include the
set of parameter values in P_b.

1.3   The Role of Parameters on Creation of Multiple Solutions

The parameters in P_b may possess a special subset

$P_{bc} \subset P_b$ in which the bifurcation phenomenon takes place. This

subset in the simplest case may correspond to a single point
$P_{bc} \in P_b$; in the most complex case to a hypersurface embedded in

P_b. This surface, as the boundary of subspaces in P_b, divides

P_b into regions. Corresponding to each of these regions, the

number of multiple solutions in X is different from that of
solutions corresponding to the neighboring regions in P_b.

Assuming that the variations in the number of multiple solutions are the result of bifurcation, by
definition P_{bc} represents the bifurcation set.

2.   THE CONCEPT OF BIFURCATION AND ITS ELEMENTS.

Critical Solutions

The system $f: (X,P) \rightarrow X$ usually possesses infinitely many solutions. Particularly, if f represents a vector field as in the case of equations of a dynamic system these solutions depend on the initial conditions and each solution becomes a trajectory of the dynamic system. However, the significant solutions in the bifurcation analysis are only the critical solutions. These are:

1. $x_o = \| X | f: x_o \rightarrow x_o \|$, singular solutions.

2. $X = \| X | f: x(0) \rightarrow x(0 + T) \|$, periodic solutions.

In the case of dynamic systems given as $dx/dt = f(X,P)$, the first set of solutions above corresponds to $dx/dt = 0$, and thus form the solution to the system of equations $f(X,P) = 0$.

Periodic Solutions

In the case of periodic solutions, the second set of critical solutions may represent <u>limit</u> <u>cycles</u>, simple or quite intricate ones.

Attractors

The more complex solutions than those above are those either almost periodic solutions or attractors on which the trajectories remain without wandering around. In short, $X_a \subset X$.

3. $X_a = \| X | f: X_a \rightarrow X_a \|$.

Dynamic Stability

The critical solutions of a system, either singular solutions, periodic solutions or complex solutions, are those solutions to which regular solutions approach either as $t \rightarrow + \infty$ or as $t \rightarrow - \infty$. For more formal description of one of the types of stability of solutions in dynamical systems, refer to Gurel and Lapidus (1969).

Structural Stability

In addition to the stability concept of solutions, we observe the concept of stability of the entire system such that as parameters change whether the two neighboring regular solutions (trajectories) remain sufficiently close to each other or not.

These are referred to as <u>structurally stable</u> and structurally unstable systems, respectively, Sotomayor (1973).

   Bifurcating Solutions

   The concept of critical solutions and their stability properties are the basic elements of the bifurcation phenomenon.
Should a generating singular point of a system split to yield
multiple critical solutions discussed above, this may well be an
example of bifurcation. However, there is more than just that to
guarantee what we observe is not merely a stability change or a
change in the number of critical solutions, instead it is an
example of bifurcation phenomenon.

   There is a further comment to be made here. The stability
of a generic singular point may be discussed in the sense of
Liapunov. Further, the structural stability of the system would
be the question of interest as is the case in bifurcation analysis.

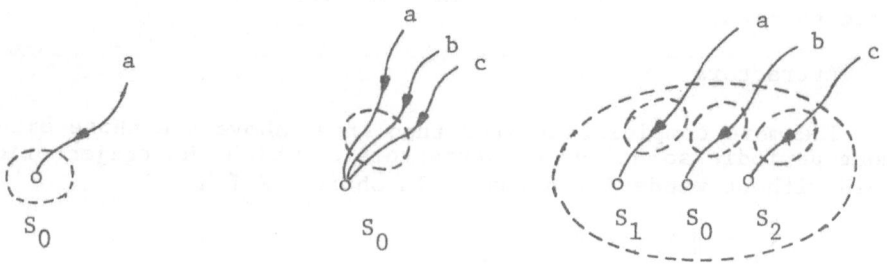

Liapunov Stability   Structural Stability   Global Stability
      $t \rightarrow \infty$ $-\infty < p < +\infty$

                                                     $a,b \rightarrow CLS = \|S_o, S_1, S_2\|$
                                  $a,b \rightarrow S_o$

   The concept of bifurcation (peeling) combines these two
stability concepts as it is related to both. If the generating
singular point S_o, stable in the sense of Liapunov, peels to
yield S_o, S_1, and S_2, although locally the system is

structurally unstable, globally the stability is observed.
Naming the set (S_o, S_1, S_2) as the Critical Limiting Set,

CLS(X), of the solution space X, the <u>Global Stability Theorem</u> is
stated as (Gurel, 1975 and 1976):

> The stability properties of the CLS(X)'s before and
> after peeling are identical.

There are some extensive books where the question of bifurcation is discussed in detail. The book by Andronov, et al,
(1973) is an account of the work initiated by A. A. Andronov
and his collaborators in the early 1930s. In fact the concept
of structural stability dates back to this period and to the
Andronov school.

A recent work where some of the modern concepts in the
bifurcation analysis are presented is by Marsden and MacCracken
(1976). This contains the introductory chapters as well as contributions by various researchers on different relevant topics.
The central theme is the bifurcations to periodic solutions where
E. Hopf's 1942 theorem is instrumental.

A recent conference on bifurcation theory and its applications resulted in a proceedings which covers diverse aspects
and applications of bifurcations (Gurel and Rossler, 1978).

Bifurcation Conditions of Poincaré at a Single Generating
Point

Poincaré not only coined the term bifurcation but also described the conditions when satisfied, a bifurcation takes place,
(Gurel, 1978a). His earlier definition in 1885 regarding the
bifurcation of $F(x_1, \ldots, x_n)$ was subsequently extended to

dynamic systems. In short, there are three conditions:

1. (Critically) Hessian must vanish at the bifurcation
    set in the parameter space.
2. (Quality) The stability change of the bifurcating point
    must accompany the first condition.
3. (Quantity) The number of critical solutions must also
    alter.

Above we discussed in general terms the multiplicity (quantity) and stability (quality) regarding the conditions 3 and 2.
These two conditions together with the first condition (criticality) yields significant information about the bifurcations
parameters and their values.

Cooperative Peeling.   Bifurcation Conditions of Multiple
Singular Solutions.

Following the arguments of global stability one can extend the Poincare conditions for peeling of a single singular
point to the conditions for that of multiple solutions.   Here
the Hessian, an n × n matrix at each singular point would be
put in a canonical form and combined to give a combined matrix,
n × r, where r is the number of the singular solutions in the
CLS(X).   The first condition is now based on this combined
Hessian.   If the CLS(X) peels cooperatively, the above three
conditions are satisfied.   Since however we refer to CLS(X)
where the global stability theorem must be satisfied to yield
another CLS'(X), the conditions above are replaced by two
conditions (Gurel, 1978b):

1)   The necessary condition for cooperative peeling is
     vanished of the combined Hessian.
2)   The sufficient condition is that CLS(X) is replaced
     by CLS'(X).

Bifurcation from Periodic Solutions.

Unlike the previous sections, the work related to bifurcations from periodic solutions is limited.   It is clear that periodic solutions (limit cycles) may bifurcate from singular points
including saddle points, i.e., singular solutions with both
stable and unstable manifolds.   Moreover, peeling of periodic
solutions yields mathematical objects of higher dimension and
complexity.   We will illustrate this by referring to a recent
analysis (Gurel, 1977) in which it is shown specifically that at
one of the bifurcation stages a limit cycle peels to yield a
toroidal surface.   An additional observation here is that the
limit cycle is also a result of a previous bifurcation and of
the stability of the second level, Gurel (1969).   A classification
of singularities as introduced in Gurel (1973), when extended to
periodic solutions, may result in even more intricate hyper-
surfaces with rich properties.

Bifurcation from Separatrices

Another class of bifurcations yielding interesting results
is bifurcations from separatrices.   In addition to various earlier studies on this topic, as one of the recent studies, we refer
to Z. Nitecki's paper in Gurel and Rossler (1978) where bifurcation question on open manifolds is discussed.

Bifurcation and Branching

In the definition of bifurcations we referred to the three elements of analysis as quantity, quality and criticality corresponding to determining the number of critical solutions, analyzing the stability of these solutions and deciding when the above changes might take place. These three elements are integral parts of the bifurcation analysis. Moreover, the global stability theorem incorporates these three parts in the bifurcation analysis. The classical buckling analysis of Euler, introduced in the early part of the eighteenth century has been extended to various engineering systems. This type of analysis may be summarized as finding the new critical solutions. In this type of analysis, parallel to the concept of structural stability, we see a reference to the implicit function theorem. However, dynamic stability does not enter into buckling analysis as an integral part of the analysis. We would like to refer to Euler's analysis as <u>branching</u> to differentiate it from <u>bifurcation</u> (peeling). In Table I a comparison of bifurcation and branching is sketched.

TABLE I.   The Elements of Bifurcation and Branching Analysis.

Elements of Analysis	Bifurcation Analysis	Branching Analysis
Quantity	Multiple solutions	Multiple solutions
Quality	Dynamic Stability	Not essential
Criticality	Structural Stability (Linearized-Hessian)	Implicit Function Theorem

In the literature there are an extensive number of studies where the creation of new solutions is predicted by proving existence theorems and determined by applying certain special calculating techniques.

As it is the case in branching to emphasize the multiplicity of the critical solutions, in some examples of chemical literature dynamic stability change of a known solution is interpreted as leading to the creation of a stable oscillatory solution. We know however that a stability change alone may not necessarily lead to the bifurcation phenomenon.

3.   SOME ILLUSTRATIVE EXAMPLES.

We might refer to some examples to illustrate certain concepts described in the above sections.

Examples of Non-Bifurcating Systems.

The three examples including the van der Pol equation
given in Gurel (1978a) do not satisfy all of the three bifurcation conditions of Poincaré, however they may be mistaken for
examples of bifurcations. They represent phenomena different
from the bifurcation phenomenon.

The Lorenz System.

As discussed in Gurel (1976) through Gurel (1978a), the
three dimensional model of Lorenz serves the purpose of illus-
trating various aspects of the bifurcation theory. First of all,
out of the three parameters only one is a bifurcation parameter.
Along this parameter, the generating singular point peels to
yield three singular points. Further, these three points cooperatively peel to result in a limiting set known as the Lorenz
attractor. For the details of the bifurcation analysis of the
Lorenz model Gurel (1978a and b) should be consulted. For the
simulation results showing various projections of the Lorenz
attractor the reader is referred to Gurel (1976).

Decomposed Partial Peeling

An example due to O. E. Rossler is shown to possess the
property of the decomposed partial peeling in which the generating singular point splits to result in two or more multiple
complementary solutions; this example is discussed in Gurel (1977).

Bifurcation to Toroidal Surfaces

In a recent study, it is found that in the three dimensional system, the generating singular point bifurcates to yield two
singular points. Further bifurcations take place inside a bounded region formed between the two singular points. In certain
subspace of the parameter space these bifurcations lead to multiple limit cycles, and in another subspace to a toroidal surface.

4.   APPLICATIONS IN SCIENTIFIC DISCIPLINES.

Engineering Systems (Structural Engineering)

The earliest applications of bifurcation analysis have
been reported in the engineering literature. The buckling

problem of rods, of columns, plates and shells led to the eigenvalue search which corresponds to finding the bifurcation values
of the parameters. The variational technique in bifurcation
theory is motivated by and applied to these problems. Among
many applied mathematicians this application is widely known and
discussed.

Physical Systems (Phase Transitions and Fluid Dynamics)

Among the physicists the bifurcation implies somewhat different interpretation. In particular, phase transitions and
problems from fluid dynamics fill the literature in applications
in physics. Partial differential equations are studied in the
analysis of bifurcation phenomena in problems leading to turbulence. However, more excitingly, E. Lorenz's equations of
turbulence are in the minds of theoreticians as well as experi-
mentalists. The significant feature of the Lorenz model is its
behavior to give an example of complex attractors, discussed
previously.

Chemical Systems (Chemical Engineering)

One of the significant motivations in the modern state of
the bifurcation analysis comes directly from the chemical systems. In addition to feedback activation and inhibition concepts of mechanical and electrical systems, here we see the role
of the reaction order which results in strong nonlinearities in
the system equations leading to interesting bifurcations. Along
these lines, problems analyzed by Prigogine (1968) should be
mentioned.

Abstract Chemical Systems (Complex Solutions)

Because of the success in chemical reaction systems there
have been models constructed to extend these studies into some
abstract chemical kinetics, notably by O. E. Rossler, (1978).
These models bridge the Lorenz model (turbulence) in physics to
similar attractors of abstract chemical systems and hopefully to
biological systems.

Biological Systems (Molecular Kinetics and Cellular Dynamics

Biological systems are complex dynamic systems. The interesting aspect of these systems is their high complexities leading
to periodic solutions and complex solutions discussed above. An
early account of bifurcations in various biochemical models is
given in Gurel (1975). Since this study, bifurcation analysis
of recent models of chemical kinetics have entered the literature. It is shown that systems without periodic solutions for

certain parameter values may bifurcate to exhibit oscillations.
The most interesting models are those of Sel'kov (1972) and his
collaborators. At the cellular level, there have been various
attempts to incorporate bifurcation analysis into cellular dynamics, notably mitotic activities of cells as well as dynamics of
a nerve cell. For example, a bifurcation analysis of the Hodgkin-
Huxley model was first presented in Gurel (1973). Following this
work, there have been subsequent papers discussing simplified
H-H models by referring to the bifurcation theory.

Ecological Interactions

Applications of bifurcation analysis can naturally be extended to the problems of interacting species. Ecodynamics,
starting with Volterra's studies, have provided nonlinear equations to characterize the behavior of the ecosystems. Due to the
nonlinearity of such equations, oscillations and bifurcations
enter naturally. We refer to Gurel and Rossler (1973) for some
discussions on ecological problems presented at the recent bifurcation conference.

Economics and Dynamics

The economic systems may also be formulated to yield equations where parameters of the system induce bifurcations in the
solution space, e.g., commodity space. The problem becomes formally an application of the bifurcation theory. S. Smale's work
along these lines gives a clear account of the relation between
the theory in general and its relevance to the formulation of the
economical dynamics. Qualitative nature of the theory appears
to be more adaptable to problems of economics than the quantitative approaches that have been widely used.

CONCLUDING REMARKS.

The concept of bifurcation is of both theoretical and applied
interest. Here a brief account of the theoretical aspects and
certain applications are presented. Specific references are to
our own work, thus various studies appearing in the literature
are not directly cited. However, a spectrum of current approaches
in theory and diverse applications of bifurcations is provided.
Therefore contributions in almost all directions are either presented by their authors or indirectly referenced in the contri-
buted papers.

The present models reveal only the beginning of a new hori-
zon in the field of global studies of complex systems. Even
then, our understanding of systems via their bifurcation analysis

goes beyond what has been accomplished by more conventional concepts. The inherent characteristic of the bifurcation theory is
that not one but all possible solutions for all possible parameter variations are considered. If a model explains all the
collected experimental data, then in addition, it provides predictions for unobserved possibilities in the system. Therefore,
it is both a descriptive and a constructive method of analysis.

REFERENCES

Andronov, A.A., Leontovich, E.A., Gordon, I.I. and Maier, A.G.
    (1973), "Theory of Bifurcations of Dynamic Systems on a
    Plane", Joh Wiley and Sons, New York.

Field, R.J. and Noyes, R.M. (1974), J. Chem. Physics 60, pp.
    1877-1884.

Gurel, O. (1978a), "Poincaré Bifurcation Analysis", In: (5).

Gurel, O. (1978b), "Necessary and Sufficient Conditions for Co-
    operative Peeling and Multiple Singular Points", In: (5).

Gurel, O. (1977), Physics Letters, 61A, pp. 219-223.

Gurel, O. (1976), "Peeling Studies of Complex Dynamic Systems",
    In: Simulation of Systems, L. Dekker (ed.), North-Holland
    Publishing Co., Amsterdam, pp. 53-58.

Gurel, O. (1975), BioSciences 7, pp. 83-91.

Gurel, O. (1975), Collective Phenomena 2, pp. 89-97.

Gurel, O. (1973), Int. J. Neuroscience 5, pp. 281-286.

Gurel, O. (1973), Math. System Theory 7, pp. 154-163.

Gurel, O. (1969), J. Franklin Institute 288, pp. 235-238.

Gurel, O. and Lapidus, L. (1969), Ind. Eng. and Chem. 61, pp.
    30-41.
Gurel, O. and Rossler, O. E. (eds.), (1978), "Bifurcation Theory
    and Applications in Scientific Disciplines", Annals of
    The New York Academy of Sciences, to appear.

Gurel, O. and Rossler, O.E. (1977), "Bifurcation to Toroidal
    Surfaces", (Preprint).

Haken, H. (1977), "Introduction to Synergetics. Nonequilibrium Phase Transitions and Self Organization in Physics, Chemistry and Biology", Springer, Berlin-Heidelberg-New York.

Marsden, J.E. and MacCracken, (eds.), (1976), "The Hopf Bifurcations and its Applications", Lectures in Applied Mathematical Science, No. 19, Springer-Verlag, New York.

Prigogine, I. and Lefever, R. (1968), J. Chem. Physics 48, pp. 1695-1700.

Sel'kov, E.E. (1972), "Nonlinearity of Multienzyme Systems", In: Analysis and Simulation of Biochemical Systems, H.C. Hemker and B. Hess (eds.), North-Holland, Amsterdam, FEBS vol. 25, pp. 145-161.

Sotomayor, J. (1973), "Structural Stability and Bifurcation Theory", In: Dynamical Systems, M.M. Peixoto (ed.), Academic Press, New York, pp. 549-560.

Thompson, J.M.T. and Hunt, G.W. (1973), "A General Theory of Elastic Stability", John-Wiley, London.

GENERAL CONSIDERATIONS OF STABILITY IN CELESTIAL MECHANICS

Victor Szebehely

The University of Texas at Austin

ABSTRACT. These lectures outline several concepts and methods of
stability concerning dynamical systems with emphasis placed on
applications to dynamical astronomy and celestial mechanics.

1.   INTRODUCTION.

The aim of celestial mechanics is to derive solutions of
the differential equations of the three and n-body problems as
well as to reveal the interrelations among these solutions or,
in other words, to describe the behavior of the dynamical system.
We must be able to treat the problems with various values of the
masses of the participating bodies and for various initial conditions. Not only are we interested in predicting the positions
and velocities of the celestial bodies for arbitrary long times,
but we must offer generally valid, qualitative results. The
performance of these tasks encounters immediate and serious difficulties. A few approaches may be described in the following.

We might attempt to establish <u>integrals</u> of the given dynamical system and to reduce its order. It has been shown that the
equations of celestial mechanics do not have a sufficient number
of uniform integrals, i.e., that the order of differential equations is higher than the number of uniform integrals. Note that
additional integrals of local validity appear often in the literature. While these are sometimes of considerable interest, they
do not violate the non-existence of sufficient number of integrals
of global validity. Another approach is to find <u>series solutions</u>
of the differential equations. This method often is referred to
as general perturbation theory. It has been shown that these

Victor G. Szebehely (ed.), Instabilities in Dynamical Systems. 61-66.
Copyright © 1979 by D. Reidel Publishing Company.

series are non-uniformly convergent and consequently, their
evaluation for arbitrary long time is meaningless. Furthermore,
the general behavior of a dynamical system cannot be ascertained
from infinite series solutions. We note the most fortunate
circumstance according to which some of these series solutions
are asymptotically convergent and the summation of a few terms
give, for a limited time-interval accurate results. This statement is of no use when we are interested in predictions for
arbitrary long time periods or when stability problems are to be
treated. The third approach to obtain solutions of celestial
mechanics is to integrate the equations of motion, numerically.
This is often referred to as special perturbation theory. It is
well known that solutions cannot be obtained this way for arbitrary long time since the accumulation of errors, sooner or later
renders the results completely meaningless. The fourth approach
is the use of topological methods based on the previously mentioned integrals of the system. This is often referred to as
the qualitative theory. Hill, Poincaré, and G.D. Birkhoff offered examples and showed that this approach might be the only
salvation for non-integrable dynamical systems.

Studying problems of stability in celestial mechanics,
we find that all the previously mentioned problems emerge immediately and most forcefully. An additional problem is that
there are so many different concepts of stability that it takes
careful consideration to establish which should be used under
given circumstances or, in fact, which definition is being used
by a given author. One way to formulate the problem of stability
of the solar system, for instance, is to find the answer to the
question of what is the interval of time at the end of which
the solar system deviates from the present configuration by a
previously assigned small amount. This is just one of the many
stability formulations which concern the solar system.

There are many side questions associated with stability. We
are interested in questions associated with possible reversibility,
and often with possible recurrence. We are concerned with the
ergodicity of the system, and with the density of periodic solutions. We would like to know if gravitational forces alone
describe the system completely, or if other effects are important.
We are interested in the possible effects of dissipation and
tidal forces, in the effect of accretion of planetary masses and
how this influences stability. We are interested in the possibilities of escapes and captures and in the effect of commensurabilities. We must study secular trends in the numerical solution
for the semi-major axis, as well as secular terms in the series
solution of the differential equations. The effects of the eccentricities and inclinations which occur in the system seem to
have crucial effects.

A recent careful listing offered over thirty different definitions and concepts of stability. Unquestionably, one of the most popular techniques to investigate stability is linearization. This method is straightforward and the results are of considerable interest and importance, especially when they reveal instabilities. The reason for this of course is that if the linearized system shows instability then the corresponding nonlinear system will also be unstable. Unfortunately, a corresponding statement does not exist when the linearized system is stable. We speak about asymptotic, conditional, weak, permanent, orbital, temporary, unilateral, numerical, secular, practical and trigonometric stability, just to mention a few. Concepts associated with the names of Birkhoff, Hill, Lagrange, Laplace, Liapunov, Poincaré and Poisson are often referred to.

2.   APPROACHES TO PROBLEMS OF STABILITY IN CELESTIAL MECHANICS.

An efficient and generally accepted method is to use the existing integrals of the dynamical system to establish boundries and limiting ranges of possible motions. Hill proposed this in 1886 for a dynamical system known as Hill's lunar problem which is a version of the restricted problem, using a simplified form of the Jacobian integral. Today we extend this and use the complete restricted problem and its Jacobian integral to study the topology of the manifold of the motion. The surfaces or curves which limit the motion (in three or two-dimensional applications) are known as the zero-velocity surfaces and curves. It is interesting to note that the geometrical details of these curves for the two-dimensional case are well known, but only sketchy treatments exist for the three-dimensional case. The singularities of the manifold correspond to the so-called equilibrium, libration or Lagrangian points. The corresponding values of the Jacobian constant are known as bifurcation values since the behavior of the system may change at these values of the Jacobian constant. The topological situation is much more complicated when the assumptions of the restricted problem are not valid and when the general problem of three or n-bodies must be taken into consideration. Recent results by Golubev (1968), Smale (1970), Saari and Marchal (1976) and Zare (1976), are of considerable interest, utilizing a combination of the momentum and energy integrals. It is essential to note that while surfaces and curves for the restricted problem bound the motion under certain circumstances, these surfaces in the general problem of three bodies are never completely closed. This means that there is always a possibility for an escape of one of the bodies. These zero-velocity curves and surfaces offer less information than other stability results; nevertheless, there are no approximations involved, and the results are exact. As an interesting application we might mention that Hill, in his original work, found that the Moon's motion must take place in a bounded region around the Earth and, consequently, neither escape nor capture is allowed. When the complete restricted problem (as opposed to Hill's approximation) is applied to the same problem, we find that the Moon's motion is still limited in a region around the Earth. Hill called the motion of the Moon stable because it cannot depart from the Earth and cannot become a satellite (or a planet) of the Sun. If the Sun-Earth-Moon system is considered as a general three-body problem, and one constructs the zero-velocity surfaces, one finds that the Moon represents a borderline case of instability and consequently it might escape or it might have been captured. This is an immediate consequence of the eccentricity of the Earth's orbit around the Sun, neglected in the model of the restricted problem.

     Another approach to problems in Celestial Mechanics as mentioned before is to use series solutions. As previously mentioned, the non-existence of uniformly convergent series certainly hinders such investigations. It is well known that the solution of the three-body problem exists as shown by Sundman but the convergence of his series solution is very slow. Along the same lines, the Kolmogorov-Arnold-Moser (1973) theory (known as the K-A-M theory) is to be mentioned. This theory requires small perturbations, does not allow the existence of low commensurabilities, requires continuous force functions and under these conditions, offers the existence of quasi-periodic motions. Unfortunately, the conditions of the K-A-M theory are not satisfied for any observed systems in Celestial Mechanics

     Another approach to the problem of stability is through numerical methods. One must be careful in applying numerical techniques, since with numerical integration, the error accumulation might introduce numerical instability and might result in a confusion and in an undesirable interaction between dynamical and numerical instability. Along these lines, it is interesting to mention the Kuiper-Nacozy-Szebehely (1977) theory (in short, the K-N-S theory) which proposes to perform numerical integration of planetary problems using exaggerated masses of the participating planetary bodies. The idea makes use of the fact that three-body problems, unless special initial conditions are used, show Laplacian instability when the participating masses are of the same order of magnitude. Considering for instance the Sun-Jupiter-Saturn problem, and increasing the masses of Saturn and Jupiter we find sudden instabilities appearing as the masses are increased above, say, twenty times their original values. The same result is obtained by analytical methods using the previously mentioned manifold of possible motions.

3.   THE HILL–LIAPUNOV METHOD.

Because of its recently revived applications, the approach
known in celestial mechanics as Hill's (1878) method or the
method of zero-velocity curves should be discussed in some detail.
(The fact that Liapunov's [1892] direct method, well known in
control theory, shows great similarity to and generalizes Hill's
method, has been pointed out to me by Dr. J. Henrard [1978].)

The advantages of this method are that it takes into consideration nonlinear effects, that it is applicable to nonintegrable dynamical systems and that its results are exact.   The
disadvantages are that the method disregards commensurabilities
which sometimes play dominant roles in stability investigations,
that the phases of the motion do not enter the determination of
the Hill curves, and that the initial conditions are replaced by
one simple parameter, which does not represent the complete
dynamical picture.   As a consequence, another disadvantage of
the method is that it only offers limiting ranges of motion.
Furthermore, even when the bifurcation values are exceeded and
the limiting curves disappear, the motion might still be stable
in a linear or nonlinear sense.

The use of the method requires the utmost care.   Consider,
for instance, the Sun–Earth–Moon system on one hand, and the
Sun–Jupiter–Saturn system on the other.   According to the method
the Moon's orbit is considered stable, if it is inside a closed
region around the Earth.   Saturn's orbit, on the other hand is
considered stable if Saturn is not allowed to penetrate the
Jupiter–Sun region.   That is, if Saturn's orbit is separated
from the region of the Sun and Jupiter.   Clearly, in the first case
the Moon might collide with the Earth, and the motion still
would be considered stable.   Instability in the Moon's case means
that the Moon escapes the Earth, and either becomes a satellite
of the Sun or escapes the system.   Instability of Saturn means
that it becomes either a satellite of Jupiter or a planet of the
Sun, inside Jupiter's orbit.

The method allows a quantitative evaluation of stability
by finding how far a stable system is removed from its critical
state.   The characteristic parameter describing the state may
be the Jacobian constant or the energy or various combinations
of the inegrals of the motion.   If the value of this parameter
for the actual stable motion is far removed from its critical or
bifurcation value, the motion is more stable, than if these values are close.   According to this we find, for instance, that
using the model of the restricted problem, the Moon's measure of
stability is of the order of $+10^{-4}$.   Using the model of the
general problem of three bodies, the stability measure becomes
-10^{-4}.   We might conclude that the Moon is marginally stable in

the first case and it is just on the unstable side in the
second. Application of the method to the satellites of the solar
system shows that all satellites are stable except the four outer
(retrograde) satellites of Jupiter when the model of the restricted problem is used.

The method may also be applied to the stability investigation
of triple stellar systems. The results show that as long as the
ratio between the semi-major axes of the outer (third) star and
the inner binary is larger than approximately 3.2, no exchange
will take place. It is of some interest to note that a similar
result has been obtained by numerical integrations. The eccentricity of the outer orbit, once again reduces the stability of
the system. The observed triple stellar systems have a stability measure of the order one. If it is compared with 10^{-4},
corresponding to the Moon, we do see that triple stellar systems
are considerably more stable than satellite systems.

4.   CONCLUSIONS

These lectures may be concluded with the general comment
that our knowledge of the stability of nonintegrable dynamical
systems occurring in Celestial Mechanics and Dynamical Astronomy
is in a somewhat elementary state. Certain special systems have
been investigated with some success; nevertheless, general questions concerning the behavior of the solar system and its long
term stability have not been answered yet. We must remember that
the physical parameters of the system might change and consequently the long term stability problem, which is the major difficulty in this field, might have to be completely reformulated.

5.   REFERENCES

Bozis, G. (1976), Astrophys. Sci., 43, 355.
Golubev, V.G. (1968), Doklady Akad. Nauk, USSR, 180, 308.
Hagihara, Y. (1957), "Stability in Celestial Mechanics," Kasai
    Publ., Tokyo.
Henrard, J., (1978), Private communication.
Hill, G. (1886), Acta Math., 8, 1.
Liapunov, A.A. (1892), Communications of the Mathematical Society
    of Krakow [2], 2, 1.
Marchal, C. and D. Saari (1975), Celestial Mechanics, 12, 115.
Moser, J. (1973), "Stable and Random Motions," Princeton Univ.
    Press, Princeton, N.J.
Smale, S. (1970), Invent. Math., 11, 45.
Szebehely, V. and R. McKenzie (1977), Astron. J., 82, 79.
Zare, K. (1977), Celestial Mechanics, 16, 35.

PART II

ASPECTS OF NUMERICAL ANALYSIS AND STATISTICAL MECHANICS

AN N-BODY INTEGRATION METHOD IN CO-MOVING COORDINATES

S. J. Aarseth

Institute of Astronomy, Cambridge, England

ABSTRACT. The Ahmad-Cohen (1973) N-body integration method is developed in co-moving coordinates with an additional time-smoothing of the type $t' = r^{3/2}$. This formulation is well suited for studying cluster formation in an expanding universe.

1. INTRODUCTION.

Direct N-body integration methods have been used for studying a variety of dynamical problems. The standard polynomial method (Wielen, 1967; Aarseth, 1972) has proved especially popular because it is easy to use and capable of dealing with several hundred interacting particles. This type of method became more powerful with the introduction of the Kustaanheimo-Stiefel regularization procedure which provides an improved treatment of critical two-body encounters and persistent binaries (Bettis and Szebehely, 1972; Aarseth, 1972). The Ahmad-Cohen (1973) scheme resulted in a further significant gain of computing efficiency, permitting somewhat larger (N $\sim$ 1000) systems to be studied. This method also lends itself naturally to KS regularization, although the programming effort is considerable (Aarseth, 1978).

The AC method has been used extensively to study cluster formation in the expanding universe, where each mass point represents a galaxy (Aarseth, Gott and Turner, 1978). This is an important new type of N-body problem which is likely to become increasingly popular. The special nature of the problem (i.e. expansion) may be exploited by a numerical formulation based on co-moving coordinates. The basic idea of the AC scheme consists

Victor G. Szebehely (ed.), Instabilities in Dynamical Systems. 69-80.
Copyright © 1979 by D. Reidel Publishing Company.

of dividing the total force acting on a particle into an irregular component due to the nearest neighbour and a distant regular
component. In the expanding universe calculation it is natural
to assign the members of one proto-cluster to the neighbour field
of a local particle; the remaining force field then behaves in a
smooth manner which can be predicted for long times.

A fairly simple co-moving neighbour scheme was developed by
Peebles and Groth (1976) and used to simulate hyperbolic cosmological models. The present forumlation is a direct application
of the AC scheme which is also suitable for more modest expansion rates (e.g., parabolic models) when the clustering process
is more pronounced. In addition to using co-moving coordinates,
a time transformation of the type $t' = r^{3/2}$ is also employed.
This feature increases the relevance of universe simulations to
the field of Celestial Mechanics and provides further stimulus
for our classical subject.

2. CO-MOVING FORMULATION.

We begin by adopting a standard soft potential
$Gm/(r^2 + \varepsilon_o^2)^{\frac{1}{2}}$ due to an extended body of mass m at a distance
r. The softening parameter ε_o may be associated with the half-
mass radius of galaxies; however, the Newtonian case if formally
recovered by setting $\varepsilon_o = 0$. The corresponding equation of
motion for one member of an N-body system is given by

$$\ddot{r}_i = -G \sum_{j \neq i}^{N} \frac{m_j(r_i - r_j)}{[|r_i - r_j|^2 + \varepsilon_o^2]^{3/2}} \tag{1}$$

Let the total mass M be distributed inside a spherical region of
space. Using Newtonian cosmology, the equation of motion for the
boundary radius R is

$$\ddot{R} = - \frac{GM}{R^2} . \tag{2}$$

We now introduce co-moving coordinates $\underline{\xi}_i$ defined by

$$\underline{\xi}_i = \frac{\underline{r}_i}{R} .$$ (3)

The new equation of motion is most readily derived by differentiating Equation (3) twice in the form $\underline{r}_i = R\underline{\xi}_i$, giving

$$\ddot{\underline{\xi}}_i = -\frac{2\dot{R}}{R}\dot{\underline{\xi}}_i + \frac{\ddot{\underline{r}}_i}{R} - \frac{\ddot{R}}{R^2}\underline{r}_i .$$ (4)

Substitution from Equations (1) and (2) together with the definition (3) results in the co-moving equation of motion

$$\ddot{\underline{\xi}}_i = -\frac{2\dot{R}}{R}\dot{\underline{\xi}}_i - \frac{G}{R^3}\left\{\sum_{j\neq i}^{N}\frac{m_j(\underline{\xi}_i - \underline{\xi}_j)}{[|\underline{\xi}_i - \underline{\xi}_j|^2 + \varepsilon^2]^{3/2}} - M\underline{\xi}_i\right\},$$ (5)

where $\varepsilon = \varepsilon_o/R$ represents the corresponding softening parameter. Alternatively ε^2 may be replaced by $\varepsilon_i{}^2 + \varepsilon_j{}^2$ where the latter values are related to the individual masses. We note that the term in brackets vanishes for a homogeneous system. The co-moving formulation is therefore well suited to systems where the peculiar motions are small compared to the expansion effect.

Equation (5) may be used in its present form (cf. Peebles and Growth, 1976). However, the presence of R^3 in the denominator of the force summation term is cumbersome in the full AC scheme which employs three explicit differentiations for the neighbour correction procedure. In order to reduce or remove this undesirable feature we introduce a time smoothing defined by the differential relation.

$$dt = R^{\nu} d\tau .$$ (6)

Application of the differentiation rule

$$\frac{d}{dt} = \frac{1}{R^{\nu}}\frac{d}{d\tau}$$ (7)

gives rise to the relations

$$\dot{\underline{\xi}}_i = \frac{1}{R^\nu} \underline{\xi}_i' \, , \tag{8}$$

$$\ddot{\underline{\xi}}_i = \frac{1}{R^{2\nu}} \underline{\xi}_i'' - \frac{\nu R'}{R^{2\nu+1}} \underline{\xi}_i' \, , \tag{9}$$

where prime denote differentiation with respect to the fictitious time τ . Using these relations, the new equation of motion
takes the general form

$$\underline{\xi}_i'' = \frac{(\nu-2)R'}{R} \underline{\xi}_i' - \frac{GR^{2\nu}}{R^3} \left\{ \sum_{j \ne i}^{N} \frac{m_j (\underline{\xi}_i - \underline{\xi}_j)}{[|\underline{\xi}_i - \underline{\xi}_j|^2 + \epsilon^2]^{3/2}} - M\underline{\xi}_i \right\} \tag{10}$$

The choice $\nu = 2$ would remove the velocity dependent term
which acts as a damping force while leaving a product involving
R linearly. This would still be cumbersome for the explicit
differentiation procedure and we therefore adopt $\nu = 3/2$ which
removes completely product terms involving a summation over N.
It may be noted that a time smoothing transformation of the type
 $t' = r^{3/2}$ (with r being the smallest two-body separation) applied to Equation (1) appears to be the best choice for the
integration of N-body systems (Szebehely and Bettis, 1972).
However, it should be emphasized that the adopted value of ν
is based entirely on practical considerations which do not apply
to a self-starting method, and in any case the role of R is different. We also note that a constant physical velocity increases
in magnitude when expressed in the new units.

The companion equation for R may be derived in a similar
way. First the relation between the two types of derivatives is
given by Equation (9) with R substituted for $\underline{\xi}_i$. The desired
equation of motion is obtained by setting $\nu = 3/2$ and substituting from Equation (2):

$$R'' = \frac{3R'^2}{2R} - GMR. \tag{11}$$

The solution for R is conveniently obtained by Taylor series. For this purpose we use two additional explicit derivatives which simplify to

$$R''' = (\frac{3R'^2}{R^2} - 4GM) R'$$ (12)

$$R^{(IV)} = 15(\frac{R'^2}{2R^2} - GM) \frac{R'^2}{R} + 4R(GM)^2 .$$ (13)

The associated integration interval $\Delta\tau_R$ is chosen conservatively in case the first term of Equation (10) makes a significant contribution.

Integration of the physical time (if required) proceeds in a similar manner. The first three derivatives take the form

$$t' = R^{3/2}$$ (14)

$$t'' = \frac{3}{2} R^{1/2} R' ,$$ (15)

$$t''' = \frac{3R'^2}{R^{1/2}} - \frac{3}{2} GMR^{3/2} .$$ (16)

For completeness, the physical coordinates and velocities are obtained by the transformations

$$\underline{r}_i = R\underline{\xi}_i ,$$ (17)

$$\underline{\dot{r}}_i = \frac{R'}{R^{3/2}} \underline{\xi}_i + \frac{1}{R^{1/2}} \underline{\xi}'_i .$$ (18)

Note that the case of no peculiar velocities (i.e., $\underline{\xi}'_i = 0$) corresponds to Hubble's law $\underline{\dot{r}}_i = H\underline{r}_i$, where $H = \dot{R}/R$ denotes the instantaneous expansion rate.

Particles which are accelerated with respect to the local Hubble flow may eventually cross the boundary at $r = R$. To simulate a balance between loss and gain across the boundary we perform a mirror reflection of any outgoing particles. The new radial velocity is then given by

$$\xi_i' \, (\text{new}) = -\xi_i' \, (\text{old}) \tag{19}$$

Note that in order to prevent spurious reflections it is necessary to employ the boundary conditions $\xi_i > 1$, $\xi_i' > 0$. Reflected particles are subjected to a complete restart (i.e., polynomical initialization) procedure to avoid discontinuity effects in the force derivatives. The boundary treatment still permits an energy interval to be used for checking purposes, provided that the resulting loss of kinetic energy is evaluated. Conversion to physical units yields

$$\Delta E_i = \frac{2m_i R'}{R^3} \, (R'\xi_i + R\xi_i' \, (\text{old}) - R'). \tag{20}$$

3. OUTLINE OF THE AC SCHEME.

The basic idea of the AC scheme is to represent the total force $\underline{F}_i$ acting on a particle as a sum of two contributions which can be treated on different time-scales. We write

$$\underline{F}_i = \underline{F}_n + \underline{F}_d \, , \tag{21}$$

where the so-called irregular component $\underline{F}_n$ is due to the nearest neighbours defined by a sphere of radius ξ_n. The velocity dependent term in Equation (10) varies on the same time-scale as the irregular force field, hence we adopt

$$\underline{F}_n = -\frac{R'}{2R}\underline{\xi}'_i - G\sum_{j\neq i}^{N}\frac{m_j(\underline{\xi}_i - \underline{\xi}_j)}{[|\underline{\xi}_i - \underline{\xi}_j|^2 + \epsilon^2]^{3/2}} , \qquad (22)$$

$$|\underline{\xi}_i - \underline{\xi}_j| < \xi_n .$$

Separate force polynomials for each force component are initialized using the principle of explicit differentiation. Thus the first derivative of the neighbour force takes the form

$$\underline{F}'_n = -\frac{R'}{2R}\underline{\xi}''_i - (\frac{R'^2}{4R^2} - \frac{1}{2}GM)\underline{\xi}'_i -$$

$$-G\sum\frac{m_j(\underline{\xi}'_i - \underline{\xi}'_j)}{[|\underline{\xi}_i - \underline{\xi}_j|^2 + \epsilon^2]^{3/2}} +$$

$$3G\sum\frac{m_j(\underline{\xi}_i - \underline{\xi}_j)[(\underline{\xi}_i - \underline{\xi}_j)\cdot(\underline{\xi}'_i - \underline{\xi}'_j)]}{[|\underline{\xi}_i - \underline{\xi}_j|^2 + \epsilon^2]^{5/2}} . \qquad (23)$$

Note that the corresponding expression for $\underline{F}'_d$ contains a term $GM\underline{\xi}'_i$ which must be retained here rather than absorbed into $\underline{F}'_n$. Two further successive differentiations are performed in a similar manner to yield $\underline{F}''_n$ and $\underline{F}'''_n$ as well as $\underline{F}''_d$ and $\underline{F}'''_d$.

Individual time-steps for both components are specified as follows. First the intervals $\Delta\tau_n$ and $\Delta\tau_d$ are obtained from simple convergence considerations by

$$\Delta\tau_n = \left[\frac{\eta_n|\underline{F}_n|}{|\underline{F}''_n|}\right]^{1/2}, \qquad (24)$$

$$\Delta\tau_d = \left[\frac{\eta_d \, |F_d|}{|F''_d|} \right]^{1/2} \tag{25}$$

The latter interval is then increased iteratively by a small numerical factor (e.g., 1.05), provided that the predicted change of the regular force does not exceed a specified fraction of the irregular force:

$$[(\tfrac{1}{6} |F'''_d| \, \Delta\tau_d + \tfrac{1}{2} |F''_d|) \, \Delta\tau_d + |F'_d|] \, \Delta\tau_d < \eta_d \, |F_n|. \tag{26}$$

The final value of $\Delta\tau_d$ is subject to the condition $\Delta\tau_d < \Delta\tau_{max}$, with $\Delta\tau_{max} = \tfrac{2}{3} R/R'$. The integration accuracy is controlled by the dimensionless parameters η_n and η_d which are usually taken to be 0.02 and 0.04, respectively.

Having specified individual integration steps, the Taylor series derivatives are converted to two sets of divided differences by the general relations

$$D_1 = (\tfrac{1}{6} F''' \, \Delta\tau_1 - \tfrac{1}{2} F'') \, \Delta\tau_1 + F' , \tag{27}$$

$$D_2 = \tfrac{1}{2} F'' - \tfrac{1}{6} F''' \, (\Delta\tau_1 + \Delta\tau_2) , \tag{28}$$

$$D_3 = \tfrac{1}{6} F''' , \tag{29}$$

where $\Delta\tau_k = \tau_o - \tau_k$. The time τ_o (which is set to zero initially) denotes the time of the most recent force calculation for each component. In addition the scheme employs the three backwards times τ_1, τ_2, τ_3 which are initialized by setting $\tau_k = -k\Delta\tau_n$, and similarly for the regular component.

The integration itself consists of a predictor to order $\underline{F}_i'''$ followed by a fourth-order corrector for both components. Changes in the neighbour field due to mass points m_j are included consistently by calculating the first three derivatives $\underline{F}_{ij}{}^{(k)}$. The corrections are performed by first converting the divided differences for each component to corresponding derivatives by

$$\underline{F}' = (\underline{D}_3 \, \Delta\tau_2 + \underline{D}_2) \, \Delta\tau_1 + \underline{D}_1 \,, \tag{30}$$

$$\underline{F}'' = 2\underline{D}_3 \, (\Delta\tau_1 + \Delta\tau_2) + 2\underline{D}_2 \,, \tag{31}$$

$$\underline{F}''' = 6\underline{D}_3 \,, \tag{32}$$

whereupon the explicit derivatives are added or subtracted. The new differences are then obtained by the relations (27) – (29). Likewise the equations of motion are advanced by a Taylor series, where the two force polynomials are combined in an appropriate manner.

The radius of the neighbour sphere, ξ_n, is determined by a strategy which attempts to optimize the total computational effort. Thus the relative frequency of the summations depends to some extent on the neighbour membership. Making n (and therefore ξ_n) too small necessitates more frequent total force summation in order to ensure an appropriate neighbour field, whereas the limit of large n leads to similar time-scales for the two polynomials. General considerations suggest that ξ_n may be obtained from the density contrast which in the co-moving formulation is defined by

$$C = \frac{n}{N\xi_n{}^3} \,. \tag{33}$$

due to the introduction of the centre of mass motion.

4.   ALGORITHM.

The sequence of steps in one integration cycle can be summarized as follows:

1.   Determine the next particle to be treated: $i = \min_j (\tau_j + \Delta\tau_j)$.

2.   Advance the solution for R if $\tau_R + \Delta\tau_R > \tau$.

3.   Predict $\underline{\xi}_j$ to order $\underline{F}_j$ (or $\underline{F}'_j$) for all neighbours.

4.   Predict $\underline{\xi}_i$, ξ'_i to order $\underline{F}'''_i$.

5.   Obtain the current neighbour force $\underline{F}_n$.

6.   Form new differences for the irregular component.

7.   Improve $\underline{\xi}_i$, ξ'_i to order $\underline{F}_i^{(IV)}$ by the corrector.

7a. Set $\underline{F}_i = \underline{F}_n + \underline{F}_d$ (τ), $\underline{F}'_i = \underline{F}'_n + \underline{F}'_d$ (τ) by extrapolating $\underline{F}_d$, $\underline{F}'_d$.

8.   Obtain $\underline{F}_n^{(new)}$ and $\underline{F}_d$, and form the new neighbour list L_{ij}.

9.   Form new differences for the regular component.

10. Improve $\underline{\xi}_i$, ξ'_i to order $\underline{F}_i^{(IV)}$ by the regular corrector.

11. Note the loss or gain of neighbours m_j.

12. Compute derivatives $\underline{F}_{ij}^{(k)}$ and correct both types of differences.

13. Modify the radius of the neighbour sphere ξ_n.

14. Set the regular integration step $\Delta\tau_d$.

15. Perform boundary reflection if $\xi_i > 1$, $\xi'_i > 0$.

16. Set the irregular integration step $\Delta\tau_n$.

17. To obtain results, predict $\underline{\xi}_j$, ξ'_j $(j = 1, \ldots, N)$ to order $\underline{F}'''_j$.

We adopt a predicted membership

$$n_p = \frac{1}{3} n_{max} (\frac{C}{5.5})^{1/2} , \qquad (34)$$

subject to $2 < n_p < \frac{2}{3}n_{max}$. At every new evaluation of $\underline{F}_d$ the neighbour sphere radius is modified according to

$$\xi_n (new) = \xi_n (old) (\frac{n_p}{n})^{1/3} . \qquad (35)$$

This stabilization strategy has the effect of including the members of a bound proto-cluster in their respective neighbour fields since a density contrast of 5.5 indicates that a self-gravitating sub-system is forming. Conversely, particles in low-density regions are assigned a small number of distant neighbours. In this way the ratio $\Sigma n_i/N$ (i.e, total average) tends to increase by a modest factor during the calculation. Experience shows that a typical average neighbour number of about 20 is sufficient for a model with N = 4000 particles, giving time-scale ratios $\Delta \tau_d/\Delta \tau_n$ approaching a factor of 20.

The present scheme can readily be modified to include the effect of galaxy collisions and mergers. These processes are due to the loss of orbital kinetic energy by inelastic encounters between two interacting galaxies. Under suitable conditions (i.e., small relative velocity and impact parameter) the two extended mass distributions would actually combine into one object. Let the masses of the two interacting particles be denoted by m_k and m_ℓ, with $k < \ell$. The merging procedure can then be summarized as follows. First we replace the old particle of mass m_k by the combined mass $m_k + m_\ell$, with corresponding coordinates and velocities obtained by the centre of mass definition. The second particle is removed by a consistent updating of variables, including a renaming of the neighbour matrix L_{ij} for all members m_j where $j \geq \ell$. The polynomial initialization follows the standard method outlined above. Finally we maintain an energy conserving scheme by making appropriate corrections

Step 7a is performed only if a new regular force summation is not required. This is determined by the condition $\tau_n + 2\Delta\tau_n < \tau_d + \Delta\tau_d$, in which case steps 8 - 14 are omitted.

The present formulation requires the following main variables for each particle: m, $\underline{\xi}_o$, $\underline{\xi}$, $\underline{\xi}_o'$, $\underline{F}$, $\underline{F}'$, $\underline{\xi}_n$, L_{ij}, together with $\Delta\tau$, τ_o, τ_1, τ_2, τ_3, $\underline{F}_o$, $\underline{D}_1$, $\underline{D}_2$, $\underline{D}_3$ for each force polynomial. Representing the neighbour matrix L_{ij} in half–word form makes a total core size requirement of $(206 + 2n_{max})$ N bytes. With $n_{max} = 50$ this permits at least N=4000 particles to be simulated on a 1400 K byte machine.

To conclude, we remark that the co-moving formulation does not readily lend itself to KS regularization and is therefore best suited for calculations with soft potentials.

5. REFERENCES

Aarseth, S.J. (1972), "Gravitational N-Body Problem," p. 373, M. Lecar (ed.), D. Reidel, Dordrecht, Holland.

Aarseth, S.J. (1978), in preparation.

Aarseth, S.J., Gott, J.R. and E.L. Turner (1978), Astrophysical Journal (to appear).

Ahmad, A. and L. Cohen (1973), J. Comp. Phys. 12, 389.

Bettis, D.G. and V. Szebehely (1972), "Gravitational N-Body Problem, p. 388, M. Lecar (ed.), D. Reidel, Dorerecht, Holland.

Peebles, P.E.J. and E. Growth (1976), Astron. and Astrophys. 53, 131.

Szebehely, V. and D.G. Bettis (1972), "Gravitational N-Body Problem," p. 136, M. Lecar (ed.), D. Reidel, Dordrecht, Holland.

Wielen, R. (1967) Veroff. Astron. Rech. Inst., No. 19, Heidelberg.

THE GENERAL THEORY OF CONSERVATIVE STABILIZATION OF THE KEPLERIAN PROBLEM

J. Baumgarte

Mechanikzentrum der Technischen Universitat
Braunschweig, Braunschweig, West Germany

ABSTRACT. By transition to the extended phase space a time transformation is proposed, having the property that the new independent variable s (fictitious time) is an angle. Four particular cases are discussed where s becomes the a) eccentric anomaly, b) mean anomaly, c) true anomaly, d) angle of the longitude
in spherical coordinates. A stabilized system of differential
equations for the Keplerian motion is obtained in every case.

1. INTRODUCTION.

Conservative stabilization means in general, that the frequencies - in the sense of transformation into action and angle
variables - are made to "a priori constants". In Baumgarte and
Stiefel (1974) and Baumgarte (1976 a, b) this problem is discussed in detail. The stabilization procedure leads to Liapunov
stability provided additional elements are introduced and considered as "a priori constants". For instance the total energy
is such an element by transition to the extended phase space.
As an example of this procedure we perform in this paper the
stabilization of the differential equations of the Keplerian
problem. For the unperturbed motion, where the energy is a constant, these claims hold true. In the perturbed case the
Liapunov stability is replaced by the so called "ε-stability"
(Baumgarte, 1976 b).

2. A GENERALIZED HAMILTONIAN VARIATIONAL FORMALISM.

The unperturbed Keplerian problem is described in cartesian

81

Victor G. Szebehely (ed.), Instabilities in Dynamical Systems. 81-93.

coordinates x, y, z and momenta p_x, p_y, p_z by the Hamiltonian

$$H = \frac{1}{2}(p_x^2 + p_y^2 + p_z^2) - \frac{K^2}{r} \; , \quad r^2 = x^2 + y^2 + z^2 \; , \qquad (2.1)$$

with K^2 the gravitational parameter. The Hamiltonian is totally degenerated. This can be seen by the fact that the orbit is closed. Total degeneration means that only one intrinsic action variable exists (and also one intrinsic angle variable). There follows also the existence of one (intrinsic) frequency depending only on the energy E in the form

$$\omega = \omega(E) = \frac{(-2E)^{3/2}}{K^2} \qquad (2.2)$$

The other two frequencies are zero. With respect to the relation $\omega = \omega(E)$ the transition to the "extended phase space" is useful, where a new independent variable s (fictitious time) is introduced.

The physical time becomes a dependent (dynamical) variable $t = x_o = q_o$ and its conjugate momentum p_o (negative total energy; $p_o = -E$) appears in the homogeneous Hamiltonian (Stiefel and Scheifele, 1971).

It is useful to perform the conservative stabilization in a generalized Hamiltonian variational formalism (compare Baumgarte, 1978). This generalized variational principle has (with $' = \frac{d}{ds}$) the form:

$$\int_{s_o}^{s_1} ds \; \{\delta[p_o \, q_o' + \sum_i p_i \, q_i'] - \sigma\delta H^* +$$

$$+ \sigma\hat{\mu}(p_o \, \delta q_o + \sum_i p_i \, \delta q_i)\} = 0 \; , \quad i = 1,2,\ldots, n \qquad (2.3)$$

In (2.3) the symbol q_i represents generalized coordinates, p_i is conjugate momentum, and

$$H^*(q_i, p_i, q_o, p_o) = \mu[H(q_i, p_i, q_o) + p_o] \text{ is the generalized (hom}$$

geneous) Hamiltonian.

We propose the following admissable variations for solving the variational problem (2.3):

$$\mu = \mu(q_i, p_i, q_o, p_o) > 0 \text{ must be varied and is called the}$$

varied scaling function, whereas $\sigma = \sigma(q_i, p_i, q_o, p_o) > 0$

and $\hat{\mu} = \hat{\mu}(q_i, p_i, q_o, p_o) > 0$ are not varied and are called

the first and the second not varied scaling function.

P_o is defined by

$$P_o = - \sum_j P_j \frac{\partial H}{\partial p_j} \tag{2.4}$$

In order to stabilize the differential equations of the Keplerian motion it is useful to consider all perturbing forces as non-conservative forces P_i.

Conservative stabilization means the choice of the varied scaling function μ in such a way, that in the case

$$\sigma = 1, \quad P_i = 0$$

the Euler equations of the variational problem (2.3) are Liapunov stable, under the hypothesis that (the constant of motion) p_o is considered as "a priori constant".

The stabilization is performed by control terms, which are produced by the fact that μ is (in contrast to σ and $\hat{\mu}$) contained in the generalized homogeneous Hamiltonian H^*.

It is shown (Baumgarte, 1976 a, b) that for any $\sigma \neq 1$ the stability behaviour of the solutions $q_i(q_o)$, $p_i(q_o)$, $P_o(q_o)$,

with $q_o = t$ (but not of the solutions $q_i(s)$, $q_o(s)$, $p_i(s)$,

$P_o(s)$) is preserved. The aim of the first non varied scaling function σ is to get an appropriate step adaption <u>independent</u> <u>of μ.</u> The second not varied scaling function $\hat{\mu}$ is obtained from μ by using the energy relation

$$H(q_i, p_i, q_o) + p_o = 0 \tag{2.5}$$

in such a way that the perturbation terms $\sigma\hat{\mu}P_i$ and $\sigma\hat{\mu}P_o$ are simplified. This handling of the perturbation terms does not influence the ε-stability (Baumgarte, 1976 b).

The variational problem (2.3) leads to the following system of differential equations:

$$q_i' = \sigma \frac{\partial H^*}{\partial p_i} = \sigma \left\{ \mu \frac{\partial H}{\partial p_i} + [H + p_o] \frac{\partial \mu}{\partial p_i} \right\}, \tag{2.6a}$$

$$q_o' = \sigma \frac{\partial H^*}{\partial p_o} = \sigma \left\{ \mu + [H + p_o] \frac{\partial \mu}{\partial p_o} \right\}, \tag{2.6b}$$

$$p_i' = -\sigma \frac{\partial H^*}{\partial q_i} + \sigma\hat{\mu}P_i = -\sigma \left\{ \mu \frac{\partial H}{\partial q_i} + [H + p_o] \frac{\partial \mu}{\partial q_i} - \hat{\mu}P_i \right\}, \tag{2.6c}$$

$$p_o' = -\sigma \frac{\partial H^*}{\partial q_o} + \sigma\hat{\mu}P_o = -\sigma \left\{ \mu \frac{\partial H}{\partial q_o} + [H + p_o] \frac{\partial \mu}{\partial q_o} - \hat{\mu}P_o \right\}. \tag{2.6d}$$

The generalized Hamiltonian

$$H^*(p_i, q_i, p_o, q_o) = \mu(H + p_o)$$

is an integral of the motion of the system (2.6) with respect to the definition (2.4) and with $\hat{\mu} = \mu$, but only under the restriction that the initial condition

$$(H + p_o) \Big|_{s=0} = 0 \tag{2.7}$$

is satisfied. In this case the value of the integral of motion $H^* = \mu(H + p_o)$ is zero. This leads with respect to $\mu > 0$ to

the energy relation (2.5) and shows that the terms in (2.6a-d) factorized by $[H + p_o]$ are control terms.

3.   THE CONSTRUCTION RULE OF CONSERVATIVE STABILIZATION.

In this paper four different possibilities of conservative stabilization of the Kepler problem are derived from the theory of action and angle variables by specifying four different scaling functions μ. For each of the four functions μ the companion functions σ and $\hat{\mu}$ are listed.

The starting point for each conservative stabilization is, as mentioned above, the unperturbed problem. In the case of Keplerian motion the Hamiltonian is given by (2.1). For the purpose of stabilization it is desired that the generalized homogeneous Hamiltonian

$$H^* = \mu(H + p_o]$$ (3.1)

(which vanishes on every solution) be the sum of action variables (compare Baumgarte and Stiefel, 1974). Then the intrinsic frequency has the value "1". There follows that the independent variable s must be an angle variable. We know the following important angle variables in the Kepler problem:
   a)   eccentric anomaly E
   b)   mean anomaly M
   c)   True anomaly φ
   d)   geographic anomaly ψ, that is the longitude in spherical coordinates r, ϑ, ψ.

With regard to the transformation into action and angle variables we write the homogeneous Kepler Hamiltonian in spherical coordinates:

$$H + p_o = \frac{1}{2} \left\{ p_r^2 + \frac{1}{r^2} \left(p_\vartheta^2 + \frac{p_\psi^2}{\cos^2 \vartheta} \right) \right\} - \frac{K^2}{r} + p_o = 0$$ (3.2)

Here we need only the well known result of the transformation into action and angle variables (Goldstein, 1974):

$$\alpha_r + \alpha_\varphi - \frac{K^2}{\sqrt{2p_o}} = 0 , \tag{3.3}$$

where

$$\alpha_\varphi = \alpha_\vartheta + \alpha_\psi = p_\varphi = \sqrt{p_\vartheta^2 + \frac{p_\psi^2}{\cos^2\vartheta}} \tag{3.4}$$

is the magnitude of the angular momentum vector. In (3.3) the expression $\left(-\dfrac{K^2}{\sqrt{2p_o}}\right)$ is also considered as an action variable.

The essential point of the conservative stabilization is that the generalized Hamiltonian $H^* = 0$ must satisfy the following conditions:

1) $H^* = 0$ contains as function of the original Hamiltonian in the form

$$H^* = \mu(H + p_o) ,$$

where $\mu > 0$ is the varied scaling function, which performs the stabilization.

2) After transformation into action and angle variables $H^* = 0$ is the sum of action variables. From this fact it follows that the value of the (intrinsic) frequency is "1" and the independent variable s is an angle.

3) Now we postulate that the independent variable s of H^* is one of the four angles E, M, φ, or ψ, which leads to four cases of conservative stabilization.

In order to understand the rule of construction of H^* it is recalled that in the trivial case $\mu = 1$, and H^* is

$$H^* = H + p_o = 0. \tag{3.5}$$

The independent variable, s is (up to a constant) the conjugate coordinate of the momentum, p_o which appears in (3.5) as an additive isolated term. This conjugate coordinate is the physical time (time coordinate). It follows that

$$t = x_o = s + \text{const.}$$

From this consideration it follows that in each of our four cases of stabilization H^* must have the form

$$H^* = \tilde{H} + \tilde{p}_o = 0 , \tag{3.6}$$

where $\tilde{p}_o$ is the conjugate of E, M, φ or ψ.

The conditions 1), 2), 3) lead to the following rule of construction in the homogeneous Hamiltonian:

$$\frac{1}{2}(p_x^2 + p_y^2 + p_z^2) - \frac{K^2}{r} + P_o = \frac{1}{2}\left\{p_r^2 + \frac{1}{r^2}\left(p_\vartheta^2 + \frac{p_\psi^2}{\cos^2\vartheta}\right)\right\}$$

$$-\frac{K^2}{r} + P_o = 0 \tag{3.7}$$

we replace in each of the four cases the momentum $\tilde{p}_o$ by $(\tilde{p}_o - H^*)$. After solving for H^* we find in every case the stabilizing scaling function μ.

In the four cases the stabilization is achieved by appropriate construction of a stabilizing function μ. Also the companion functions σ, $\hat{\mu}$ are indicated.

4.   THE FOUR CASES OF CONSERVATIVE STABILIZATION.

a)   The independent variable s is the eccentric anomaly E, thus $s = E + \text{const.}$ The conjugate momentum of E is $\left(-\dfrac{K^2}{\sqrt{2p_o}}\right)$.

We replace in (3.7)

$$\left(-\frac{K^2}{\sqrt{2p_o}}\right) \quad \text{by} \quad \left(-\frac{K^2}{\sqrt{2p_o}} - H^*\right) \tag{4.1}$$

in such a way, that we multiply (4.1) by $(-\sqrt{2p_o})$ and obtain as the replacement of

$$(K^2) \quad \text{by} \quad (K^2 + H^* \sqrt{2p_o}) . \tag{4.2}$$

By doing this in (3.7) we find (using cartesian coordinates)

$$\frac{1}{2} \underline{p}^2 - \frac{K^2 + H^* \sqrt{2p_o}}{r} + p_o = 0 , \quad \underline{p}^2 = p_x^2 + p_y^2 + p_z^2 ,$$

$$r^2 = x^2 + y^2 + z^2 . \tag{4.3}$$

Solving for H^* we obtain

$$H^* = \frac{r}{\sqrt{2p_o}} \left[\frac{1}{2} \underline{p}^2 - \frac{K^2}{r} + p_o \right] = \frac{r}{\sqrt{2p_o}} \left[H + p_o \right] , \tag{4.4}$$

leading to

$$\mu = \frac{r}{\sqrt{2p_o}}$$

as a well known result (compare p.e. Stiefel and Scheifele, 1971 or Baumgarte, 1978).

For σ and $\hat{\mu}$ we may choose

$$\sigma = c \sqrt{2p_o} \, r^{m-1} , \quad 1 \leq m \leq 2 , \tag{4.5}$$

when c is a scaling constant and

$$\mu = \hat{\mu} = \frac{r}{\sqrt{2p_o}} . \tag{4.6}$$

b)   The independent variable is the mean anomaly M, thus
s = M + const.   The conjugate momentum of M is also $(-\dfrac{K^2}{\sqrt{2p_0}})$.

We replace in (3.7) once more

$$\left(-\frac{K^2}{\sqrt{2p_0}}\right) \quad \text{by} \quad \left(-\frac{K^2}{\sqrt{2p_0}} - H^*\right) \tag{4.7}$$

but reformulate (4.7) now by replacing

$$\left(\frac{K^2}{\sqrt{2p_0}}\right)^{-1} \quad \text{by} \quad \left(\frac{K^2}{\sqrt{2p_0}} + H^*\right)^{-1} \tag{4.8a}$$

or

$$\left(\sqrt{2p_0}\right) \quad \text{by} \quad \left(\frac{1}{\dfrac{1}{\sqrt{p_0}} + \dfrac{\sqrt{2}}{K^2} H^*}\right) \tag{4.8b}$$

and therefore

$$p_0 \quad \text{by} \quad \frac{1}{\left[\dfrac{1}{\sqrt{2p_0}} + \dfrac{\sqrt{2}}{K^2} H^*\right]^2} \tag{4.8c}$$

Consequently we replace

$$\left\{H + p_0 = 0\right\} \quad \text{by} \quad \left\{H + \frac{1}{\left[\dfrac{1}{\sqrt{p_0}} + \dfrac{\sqrt{2}}{K^2} H^*\right]^2} = 0\right\} . \tag{4.9c}$$

This leads to

$$\frac{1}{\sqrt{p_0}} + \frac{\sqrt{2}}{K^2} H^* = \frac{1}{\sqrt{-H}}$$

or

$$H^* = \frac{K^2}{\sqrt{-2H}} - \frac{K^2}{\sqrt{2p_0}} . \tag{4.10}$$

From (4.10) we calculate μ :

$$\mu = \frac{K^2}{P_o\sqrt{-2H} - H\sqrt{2p_o}} \quad . \tag{4.11}$$

It is useful to choose

$$\sigma = c \frac{(-2H)^{3/2}}{K^2} r^m , \quad 1 \leq m \leq 2 \tag{4.12}$$

and

$$\hat{\mu} = \frac{K^2}{(-2H)^{3/2}} \tag{4.13}$$

Using (4.11), (4.12), (4.13) our variational problem
(2.6a-d) leads to the following equations of motion in cartesian
coordinates

$$x_i' = \frac{\partial H}{\partial p_i} c r^m \tag{4.14a}$$

$$p_i' = \left(- \frac{\partial H}{\partial x_i} + P_i\right) c r^m \tag{4.14b}$$

$$x_o' = \left(\frac{-H}{P_o}\right)^{3/2} c r^m \tag{4.14c}$$

$$p_o' = \left(- \frac{\partial H}{\partial x_o} + P_o\right) c r^m \tag{4.14d}$$

We see that in the system (4.14a-d) only (4.14c) is modified by a control factor, which is "1" in the analytical (but
not in the computed) solution. This method seems to be the most
useful for a conservative stabilization (Baumgarte, 1976a,b).

In order to discuss the cases c) and d) it is necessary to
use spherical coordinates.

c) The independent variable s is the true anomaly φ , thus
s = φ + const. The conjugate momentum of φ is

$$p_\varphi = \sqrt{p_\vartheta^2 + \frac{p_\psi^2}{\cos^2\vartheta}} \quad .$$

We replace (in the case $p_\varphi = \sqrt{p_\vartheta^2 + \dfrac{p_\psi^2}{\cos^2\vartheta}} \neq 0$) in (3.7)

$$\left(\sqrt{p_\vartheta^2 + \frac{p_\psi^2}{\cos^2\vartheta}} \right) \text{by} \quad \left(\sqrt{p_\vartheta^2 + \frac{p_\psi^2}{\cos^2\vartheta}} - H^* \right) \qquad (4.15a)$$

or

$$\left(p_\vartheta^2 + \frac{p_\psi^2}{\cos^2\vartheta} \right) \text{by} \quad \left(\sqrt{p_\vartheta^2 + \frac{p_\psi^2}{\cos^2\vartheta}} - H^* \right)^2 \qquad (4.15b)$$

and solve (3.7) for H^* . We obtain

$$H^* = - r \sqrt{2(\frac{K^2}{r} - P_o - \frac{1}{2} p_r^2)} + \sqrt{p_\vartheta^2 + \frac{p_\psi^2}{\cos^2\vartheta}} , \qquad (4.16)$$

where the scaling function

$$\mu = \frac{r^2}{\frac{1}{2}\sqrt{p_\vartheta^2 + \frac{p_\psi^2}{\cos^2\vartheta}} + r \sqrt{\frac{K^2}{r} - P_o - \frac{1}{2} p_r^2} \frac{1}{2}} \qquad (4.17)$$

is used. The introduction of the Delaunay-similar elements with the true anomaly as the independent variable (DS φ) by Scheifele is based on a similar idea, (Scheifele, 1970: Scheifele and Graf, 1974).

For σ and $\hat{\mu}$ we may choose

$$\sigma = 1 \quad \text{or} \quad \sigma = c \sqrt{p_r^2 + \frac{p_\psi^2}{\cos^2 \vartheta}} \; r^{m-2} \;, \quad 1 \le m \le 2 \qquad (4.18a,b)$$

and

$$\hat{\mu} = \frac{r^2}{\sqrt{p_r^2 + \dfrac{p_\psi^2}{\cos^2 \vartheta}}}$$

d) The independent variable s is finally the angle ψ, hence $s = \psi + \text{const.}$ The conjugate momentum of ψ is p_ψ. We replace (for a positive p_ψ) in (3.7)

$$(p_\psi) \quad \text{by} \quad (p_\psi - H^*) \qquad (4.20)$$

and solve (3.7) for H^* :

$$H^* = - \sqrt{\{(\frac{K^2}{r} - P_o - \frac{1}{2} p_r^2) \, 2r^2 - p_r^2\} \cos^2 \vartheta} + p_\psi \;. (4.21)$$

where the scaling function

$$\mu = \frac{2r^2 \cos^2 \vartheta}{p_\psi + \sqrt{\{(\frac{K^2}{r} - P_o - \frac{1}{2} p_r^2) \, 2r^2 - p_r^2\} \cos^2 \vartheta}} \qquad (4.22)$$

is used.

For σ and $\hat{\mu}$ we may choose

$$\sigma = 1 \quad \text{or} \quad \sigma = c \frac{p_\psi}{\cos^2 \vartheta} \; r^{m-2} \;, \quad 1 \le m \le 2 \qquad (4.23a,b)$$

and

$$\hat{\mu} = \frac{r^2 \cos^2 \vartheta}{p_\psi} \;. \qquad (4.24)$$

In the cases c) and d) we may also use cartesian coordinates by performing a canonical transformation from spherical into cartesian coordinates in the variational problem (2.3). This variational formalism, indeed, is invariant with respect to canonical transformations.

Remark: If we prefer to distinguish between true perturbing non-canonical forces and forces stemming from a perturbing potential $V = V(x, y, z, x_o) = V(r, \vartheta, \psi, x_o)$, $x_o = q_o = t$, we have to replace in the Hamiltonian (2.1) the potential

$$\left(-\frac{K^2}{r}\right) \quad \text{by} \quad \left(-\frac{K^2}{r} + V\right) \tag{4.25}$$

and use the symbols P_i as the true non-canonical forces.

5.   ACKNOWLEDGEMENTS.

The author acknowledges gratefully, the support of the "Deutsche Forschungsgemeinschaft". He wishes to thank W. von Grünhagen for his helpful discussions.

6.   REFERENCES.

Baumgarte, J. (1978), "Dynamics of Planets and Satellites and Theories of Their Motion", V. Szebehely (ed.), p. 149, D. Reidel, Dordrecht, Holland.

Baumgarte, J. (1976a), Celestial Mech. 13, p. 105-109.

Baumgarte, J. (1976b), "long-Time Predictions in Dynamics," V. Szebehely and B.D. Tapley (eds.), P. 153, D. Reidel, Dordrecht, Holland.

Baumgarte, J. and Stiefel, E.L. (1974), Celestial Mech. 10, p. 71-85.

Goldstein, H. (1974), "Klassische Mechanik", Akademische Verlagsgesellschaft, Frankfurt am Main, 3. Auflage.

Scheifele, G. (1970), Celestial Mech. 2, p. 296.

Scheifele, G. and Graf, O. (1974), AIAA Paper No. 74-838, Anaheim, Calif.

Stiefel, E.L. and Scheifele, G. (1971), "Linear and Regular Celestial Mechanics", Springer-Verlag, Berlin.

CELESTIAL MECHANICS, QUANTUM MECHANICS, AND PATH INTEGRATION

Cécile DeWitt-Morette

Department of Astronomy and Center for Relativity
The University of Texas at Austin

ABSTRACT. Problems of celestial mechanics and quantum mechanics
which can be stated in terms of the same concepts and solved by
the same equations are presented.

1.   INTRODUCTION.

Discovering deep underlying connections between topics which
have entirely different starting points is one of the most exciting experiences in physics and in mathematics.   In these lectures
I shall discuss some aspects of the connections between

Stochastic systems	and	deterministic systems
Path integrals	and	differential equations
Quantum mechanics	and	classical mechanics

In their preface to Problèmes Ergodiques de la Mécanique Classique, Arnold and Avez have indicated how disciplines as varied as
probability, topology, number theory, differential geometry, etc.
enter the study of dynamical systems.   The beautiful article of
G. Mackey, "Ergodic Theory and its Significance for Statistical
Mechanics and Probability Theory" gives the mathematical reasons
underlying these connections.   These lectures will not explore
all these issues and I shall restrict myself to a fascinating development of recent years: functional integration, also called
path integration.

Differential equations and path integrals are complementary
mathematical descriptions of physical systems: the world is

Victor G. Szebehely (ed.), Instabilities in Dynamical Systems. 95-101.
Copyright © 1979 by D. Reidel Publishing Company.

global and stochastic and physical laws are local and deterministic. Differential equations are the natural formulation of
local deterministic physical laws, path integrals are the natural
formulation of global stochastic systems. The marvelous thing is
that path integrals are solutions of partial differential equations.

Since classical mechanics is the limit of quantum mechanics
when Planck's constant h tends to zero, one can use the path integral representation of the solutions of the Schrödinger equation in the discussion of classical systems - provided one knows
how to compute path integrals. I shall first give a method for
computing path integrals. Then I shall give a precise meaning
to the statement "classical physics is the limit of quantum mechanics," and finally relate some properties of quantum systems
to properties of the corresponding classical systems.

2.   FUNCTIONAL INTEGRATION.

A functional integral is an integral over a function space,
i.e., an integral over an infinite dimensional space. Integrals
over infinite dimensional spaces cannot be defined simply by a
formal generalization of integrals over finite dimensional spaces.

Path integration has been introduced in statistical mechanics
by Wiener in the early twenties, and reinvented in the early
forties by Feynman in quantum physics. Differential equations
were then such a well developed and powerful tool that it was
difficult to appreciate path integrals in their infancy. Nevertheless, they have entered in nearly all branches of physics and
there are now methods for computing them explicitly. I shall
present such a method on a famous example: How the Wiener integral accounts for the properties of Brownian motion.

1. __Domain of integration.__ Let $T = [t_a, t_b]$, let X be the
space of continuous paths x: $T \to \mathbb{R}$ such that $x(t_a) = 0$. A
continuous path - as opposed to a smooth path - is a random path:
If the path x has no derivative, i.e., if we do not know $\dot{x}(t)$, we
cannot say in which direction the particle known to be at x(t) at
time t will proceed. Hence X is the suitable domain of integration for the diffusion of particles starting at 0 at t_a.

2. __"Measure" on X.__ In general one cannot define a measure
on an infinite dimensional space. Consider for instance the
Gaussian measure on $\mathbb{R}^n$ of covariance A

$$d\gamma(u) = (2\pi)^{-n/2} \exp(-A_{ij}^{-1} u^i u^j / 2)(\det A_{ij}^{-1})^{1/2} du^1 \dots du^n.$$

One cannot generalize this to infinite dimensional spaces X unless one defines σ-fields of subsets of X. This has been done by probabilists and they have developed a beautiful theory of functional integration. Unfortunately their approach, motivated by the Wiener integral, does not work for the Feynman integral. I shall use another approach which works for both types and which, in addition, leads to a versatile technology for the computation of functional integrals. It is based on the following remark: Whereas the Gaussian measure γ is a set function, i.e., a function defined on subsets of $\mathbb{R}^n$,

$$\gamma(U) = \int_U d\gamma(u),$$

its Fourier transform

$$F\gamma(\xi) = \exp\left(- A^{ij} \xi_i \xi_j / 2\right)$$

is an ordinary function, defined pointwise, which can easily be generalized to infinite dimensional spaces. Note first that if γ is defined on a space X (here $\mathbb{R}^n$), its Fourier transform is defined on its dual X' (here $\mathbb{R}^n$). Indeed

$$F\gamma(\xi) = \int_{\mathbb{R}^n} \exp\left(- i \langle \xi, u \rangle\right) d\gamma(u)$$

the duality on $\mathbb{R}^n$ being $\langle \xi, u \rangle = \sum \xi_i u^i \in \mathbb{R}$. On the space X of continuous functions x, the duality between X and X' is given by

$$\langle \mu, x \rangle = \int_{\mathbf{T}} d\mu(t) \, x(t) \in \mathbb{R} \quad \text{for every } x \in X \text{ and } \mu \in X'.$$

Thus the Fourier transform of a Gaussian w on X is

$$Fw(\mu) = \exp\left(- \frac{1}{2} \int_{\mathbf{T}} d\mu(t) \int_{\mathbf{T}} d\mu(s) \ G(t,s)\right) \equiv \exp\left(- \frac{1}{2} W(\mu,\mu)\right)$$

The Gaussian w is characterized by the covariance G or, equivalently by the variance W. For instance the covariance of the Wiener Gaussian is

$$G(t,s) = \inf(t - t_a, s - t_a).$$

3.   A functional integral
with respect to the Wiener Gaussian w.

We shall compute

$$I = \int_X f(x(t_1) - x(t_a),\ x(t_2) - x(t_1),\ \cdots x(t_n) - x(t_{n-1}))\ dw(x)$$

and obtain

$$I = \int_{R^n} f(u^1,\ \ u^n)\ d\gamma_1(u^1) \cdots d\gamma_n(u^n),$$

where

$$u^k = x(t_k) - x(t_{k-1})$$

and where

$$d\gamma_k(u^k) = (2\Pi(t_k - t_{k-1}))^{-1/2} \exp(-(u^k)^2/2(t_k - t_{k-1}))du^k.$$

We conclude that the random variables $\{x(t_k) - x(t_{k-1})\}$ are independent for the Wiener Gaussian w for any time partition. $x(t_k)$
$-x(t_{k-1})$ has a Gaussian distribution of mean 0 and covariance
$t_k - t_{k-1}$.

First we must define the integral I with respect to a "measure" w known only by its Fourier transform. The following discussion provides both a definition of I and a method for computing it. Let P be a linear continuous mapping $P:X \to U$. Let w be
a "measure" on X and w the "measure" on Y induced by P. Let $\tilde{P}$
be the transposed mapping between the respective duals $\tilde{P}:Y' \to U'$

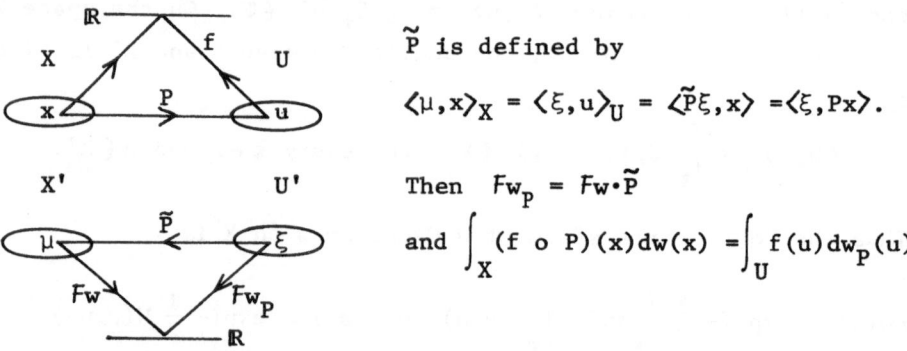

$\tilde{P}$ is defined by

$$\langle \mu, x \rangle_X = \langle \xi, u \rangle_U = \langle \tilde{P}\xi, x \rangle = \langle \xi, Px \rangle.$$

Then $Fw_P = Fw \cdot \tilde{P}$

and $\int_X (f \circ P)(x) dw(x) = \int_U f(u) dw_P(u)$

If U is a finite dimensional space and if the integral over U is
defined, this equation serves to define the integral over X. The
definition of the integral over X when these conditions are not
fulfilled, i.e., when the integral is not a so called cylindrical
integral, are beyond the scope of this lecture. In the present
example

$$P: x \to u = \{u^1, \cdots u^n\} \quad \text{where} \quad u^k = \langle \delta_{t_k} - \delta_{t_{k-1}}, x \rangle.$$

$$\langle \xi, u \rangle_U = \sum \xi_k u^k = \sum \langle \xi_k (\delta_{t_k} - \delta_{t_{k-1}}), x \rangle \quad , \text{hence}$$

$$\tilde{P}\xi = \sum \xi_k (\delta_{t_k} - \delta_{t_k}).$$

$$Fw_P(\xi) = Fw \circ \tilde{P}\xi = \exp\left(-\frac{1}{2} W (\tilde{P}\xi, \tilde{P}\xi)\right)$$

$$W(\tilde{P}\xi, \tilde{P}\xi) = \sum \xi_i \xi_j W(\delta_{t_i} - \delta_{t_{i-1}}, \delta_{t_j} - \delta_{t_{j-1}})$$

$$= \sum \xi_i \xi_j (G(t_i, t_j) - G(t_i, t_{j-1})$$

$$- G(t_{i-1}, t_j) + G(t_{i-1}, t_{j-1}))$$

$$= \sum \xi_i^2 (t_i - t_{i-1})$$

It follows that $dw_P(u) = d\gamma_1 u^1) \cdots d\gamma_n(u^n)$.

3. CLASSICAL PHYSICS AS THE UNIT OF QUANTUM PHYSICS.

Let M be the configuration space of a dynamical system, let C_t be its classical flow.

$$C_t : M \to M \quad \text{by} \quad a \to q(t, a, v_a),$$

where q is the classical path such that

$$q(t_a, a, v_a) = a \quad \text{and} \quad \dot{q}(t_a, a, v_a) = v_a.$$

In the limit h = 0, the probability of finding in ΩCM at time t the system known to be in $C_t^{-1}(\Omega)$ at time t_a is unity. This says that in the limit $\hbar = 0$ the probability of finding the dynamical system is peaked along the classical path. The classical path from (a, t_a) to (b, t_b) can be considered as an "equilibrium point" in the space of all possible paths $f : T \to M$ from (a, t_a) to (b, t_b) reached when the action $S(f)/\hbar$ tends to infinity. The classical path q is a critical point of the action, $S'(q) = 0$. The second

variation $S''(q)xx$ gives both the qualitative features of the
quantum system which in the limit $\hbar = 0$ follows the classical
path q and the stability properties of the classical system along
q. This is the reason why celestial mechanics and quantum mechanics can trade theorems and results.

The first relationship between celestial mechanics and
quantum mechanics dates back to the Einstein-Bohr-Sommerfeld
equation

$$\oint p(E_n,q)dq = n\hbar \qquad n = 1,2,\ldots$$

which gives the energy levels E_n of a quantum system in terms of
the integral $\oint p(E,q)dq$ along periodic orbits of the corresponding classical system. The bridge between classical and quantum
mechanics was built by Wentzel, Kramers and Brillouin. They
used techniques pioneered by Jeffereys and the limit when h tends
to zero of quantum mechanicsl results is known as the WKB approximation or the JWKB approximation. The WKB approximation is <u>not</u> a
classical result but a quantum result which can be computed in
terms of classical quantities. For instance the exact probability
amplitude $K(b,t_b; a,t_a)$ for a transition from (a,t_a) to (b,t_b) can
be written as

$$K = K_{WKB} \left(1 + \sum_{n=1}^{\infty} \hbar^n A_n\right).$$

K can be expressed as a path integral known as the Feynman-Kac
formula. Its WKB approximation is readily computed and the terms
A_n are given by finite dimensional integrals, following the method
outlined in Section 2.

The energy spectrum can be obtained similarly. At the WKB
approximation, one obtains the corrected Bohr-Sommerfeld equation

$$\oint p_\alpha(E,q)dq^\alpha + h\lambda/4 - i\frac{\hbar}{2\pi} \sum_{k=2}^{n} \left(m_k + \frac{1}{2}\right) \alpha_k(E) \ (E) = n\hbar,$$

where λ is an integer sum of the Morse index and the number of
"turning points" or "libration points," $\alpha_k(E)$ are the Poincaré
characteristic exponents, also called stability angles, n and
m_2, m_3, $\ldots$ m_n are quantum numbers. The solution E of the corrected Bohr-Sommerfeld equation corresponding to a particular
choice of $m_1, m_2, \ldots m_n$ should thus be labeled $E_{n,m_1,\ldots m_n}$.

Classically, a periodic orbit is linearly stable (Poincaré stability) if an only if all its characteristic exponents are purely imaginary. Quantum mechanically $E_{n,m_1,\ldots m_n}$ is real if and only if all the characteristics are purely imaginary. If not, $E_{n,m_1,\ldots m_n}$ are complex. Their imaginary parts contribute to the line-width of the energy spectra. The corrected Bohr-Sommerfeld equation has been obtained very crudely by treating quantum mechanically a particle in a potential. It does not incorporate the effects of the electromagnetic field. Nevertheless, it shows how the characteristic exponents of celestial mechanics enter the energy spectrum and the decay rates of a bounded quantum mechanical system. This is only one of several examples where concepts developed for celestial mechanics can be used profitably in quantum mechanics. The WKB bridge has been used extensively for the benefit of quantum mechanics. There is no reason why it cannot be used for the benefit of classical mechanics. It is natural that the path integral formulation of quantum mechanics which sums over all possible paths "feels" the critical points in the space of paths, i.e., the classical paths.

4. REFERENCES.

The problems presented here are developed in a monograph: Path Integration in NonRelativistic Quantum Mechanics by C. DeWitt-Morette, A. Maheshwari, and B. Nelson. It will appear in 1979 as an issue of Physics Reports. All relevant references can be found in this monograph.

STATISTICAL VIEWPOINT IN CLASSICAL MECHANICS

G. C. Stey

Instituts Internationaux de Physique et de Chimie
(Solvay), Université Libre de Bruxelles, Brussels, Belgiur

ABSTRACT. A formulation of dynamics is presented from the point
of view of statistical physics in which probability distributions,
rather than trajectories, are the principal objects of interest.
A central role is played by concepts introduced in statistical
mechanics – in particular, by the collision operator, $\psi(z)$, in
the description of the long-time behavior. This viewpoint is
illustrated by a soluble classical example.

We consider a classical mechanical system with n degrees
of freedom whose motion is generated by a Hamiltonian $H(x,p)$,
$(x,p) = (x_1 \cdots x_n, p_1 \cdots p_n)$:

$$\frac{dx}{dt} = \frac{\partial H(x,p)}{\partial p}, \quad \frac{dp}{dt} = -\frac{\partial H}{\partial x}(x,p). \tag{1}$$

The solution $(x(t), p(t))$ gives the evolution of the state of
the system as a trajectory in 2n-dimensional phase space, which
is normally determined once the initial state $(x(0), p(0))$ is
specified. The utility of the viewpoint in which individual trajectories are the fundamental objects of interest can, however,
be limited. For example in some situations the initial state is
not precisely known and the long-time behavior of the system is
strongly dependent on the initial state (Ruelle, 1977). In these
cases, infinite sets of initial states and trajectories need to
be considered. This is done in statistical physics by use of
statistical ensembles.

The density of ensemble members in state (x,p) in the

Victor G. Szebehely (ed.), Instabilities in Dynamical Systems. 103-112.
Copyright © 1979 by D. Reidel Publishing Company.

2n-dimensional phase space at time t is $\rho(x,p,t)$, which obeys the local conservation equation known as the Liouville equation

$$\frac{\partial \rho}{\partial t} = -iL\rho(x,p,t), \tag{2}$$

where

$$L = -i \sum_{j=1}^{n} \left(\frac{\partial H}{\partial p_j} \frac{\partial}{\partial x_j} - \frac{\partial H}{\partial x_j} \frac{\partial}{\partial p_j}\right) \tag{3}$$

is the Liouvillian. The solution of Equation (2), expressed formally as

$$\rho(x,p,t) = e^{-itL} \rho(x,p,0)$$

is usually studied in statistical mechanics by methods developed with large systems in mind, (i.e., systems with large number of degrees of freedom). However, some of these ideas and results apply also to systems with only a few degrees of freedom. In the following, we discuss some relevant developments in this area, obtained in the framework of the theory of non-equilibrium statistical physics of I. Prigogine and co-workers (Balescu, 1975; Prigogine, 1962; Rae, 1969).

Solution of Equation (2) by the technique of Laplace transforms gives

$$\rho(x,p,t) = \frac{1}{2\pi i} \int_B dz \ e^{-izt} (z-L)^{-1} \rho(x,p,0), \tag{4}$$

where B is a line parallel to the real z-axis, running from right to left, above all singularities of the integrand. It is convenient to introduce a projector P and its complement $Q = 1-P$; $QP = PQ = 0$. With these, the resolvent of L can be written as (Rae, 1973)

$$(z-L)^{-1} = (P + C(z)) \ a(z) \ (P + \mathcal{D}(z)) + I(z) \tag{5}$$

since

$$P(z-L)^{-1} P = a(z) = (z-\psi(z))^{-1}P \tag{6}$$

$$P(z-L)^{-1} Q = a(z) \ \mathcal{D}(z) \tag{7}$$

$$Q(z-L)^{-1} P = C(z) \ a(z) \tag{8}$$

$$Q(z-L)^{-1} Q = C(z) \ a(z) \ \mathcal{D}(z) + I(z) \ , \tag{9}$$

where
$$\psi(z) = PLP + PLQ\ (z-QLQ)^{-1}\ QLP \tag{10}$$

$$C(z) = I(z)\ QLP \tag{11}$$

$$D(z) = PLQ\ I(z) \tag{12}$$

$$I(z) = Q(z-QLQ)^{-1}\ Q \tag{13}$$

are the collision, creation, destruction and propagation operators respectively (Prigogine, 1962).

The behavior of ρ as $t \rightarrow \infty$ depends on the nature of
the spectrum of L. The long-time properties have been described
in terms of kinetic equations in which the asymptotic collision
operator plays an important role (Prigogine and Résibois, 1961).
The density-in-phase $\rho(x,p,t)$ does not tend usually to a limit as
$t \rightarrow \infty$. On the other hand, the limit of the time-average

$$\lim_{T \rightarrow \infty} \frac{1}{T} \int_0^T dt\ e^{-itL}\ \rho(x,p,0) = E\ \rho(x,p,0)\ , \tag{14}$$

where
$$E = \lim_{z \rightarrow i0^+} z(z-L)^{-1} \tag{15}$$

is the projector on to $N(L)$, the null space of L, the subspace of the dynamical invariants ϕ and $L\phi = 0$.

That the value of $\lim_{z \rightarrow i0^+} \psi(z)$ determines the nature of the
time average of $P\rho$ rather than that of $P\rho$ for $t \rightarrow \infty$ was
discovered by Stey (1974), in finding the necessary and sufficient condition that $\psi(z) \rightarrow 0$ when $z \rightarrow 0$ in systems for
which $a(z)$ and $\psi(z)$ have only poles as singularities at
$z = 0$. The following statements were found to be equivalent
(Stey, 1974):

$$\lim_{z \rightarrow i0^+} \psi(z) = 0 \tag{16a}$$

$$PEP\ f = 0\ \text{ implies }\ Pf = 0 \tag{16b}$$

$$EPf = 0\ \text{ implies }\ Pf = 0 \tag{16c}$$

$$QLf = 0\ \text{ implies }\ Lf = 0 \tag{16d}$$

$$R(PE) = R(P) \ , \quad (R = \text{range}).\tag{16e}$$

Here also, the condition $\lim\limits_{z \to i0^+} \psi(z) \neq 0$ was introduced

(Stey, 1974), as an irreversibility criterion for the relation between the limit $t \to \infty$ and the limit $t \to 0^+$ of time averages $t^{-1} \int_0^t d\tau \ f(\tau)$. For example, from the equivalence of (16a) and (16b) above, we have the following theorem:

If we write

$$Pf = \lim_{T \to \infty} \frac{1}{T} \int_0^T dt \ P \ e^{-itL} \ P \ g = PEPg\tag{17}$$

then, to a given Pf there corresponds one and only one Pg, if and only if $\lim\limits_{z \to i0^+} \psi(z) = 0$.

In the case of Hamiltonian systems described by angle-action variables and for which there exists a canonical transformation $\alpha_1 = \alpha(\tilde{\alpha}_1, \tilde{J}_1)$, $J_1 = J_1(\tilde{\alpha}_1, \tilde{J}_1)$ such that $H(\alpha_1, J_1) = \tilde{H}(\tilde{J}_1)$, independent of $\tilde{\alpha}_1$, we have

$$\langle \tilde{\alpha}_1 \tilde{J}_1 | L | \tilde{\alpha}_2 \tilde{J}_2 \rangle = -i \ \tilde{\omega}(\tilde{J}_1) \cdot \frac{\partial}{\partial \tilde{\alpha}_1} \ \delta(\tilde{\alpha}_1 - \tilde{\alpha}_2) \ \delta(\tilde{J}_1 - \tilde{J}_2).\tag{18}$$

The change of basis $\langle \tilde{\alpha}_1 \tilde{J}_1 | \nu_1 \tilde{J}_1' \rangle = \delta(\tilde{J}_1 - \tilde{J}_1') \ e^{i\nu_1 \tilde{\alpha}_1} / (\sqrt{2\pi})^n$

gives

$$\langle \nu_1 \tilde{J}_1 | L | \nu_2 \tilde{J}_2 \rangle = \nu_1 \cdot \tilde{\omega}(\tilde{J}_1) \ \delta(\tilde{J}_1 - \tilde{J}_2) \ \delta^{Kr}(\nu_1 - \nu_2) \ ,\tag{19}$$

where ν_1 is an n-dimensional vector whose components are integers and $\tilde{\omega}_j(\tilde{J}_1) = \delta\tilde{H}(\tilde{J}_1)/\delta\tilde{J}_{1j}$ $(j=1,2,\ldots n)$.
Thus,

$$\langle \tilde{\alpha}_1 \tilde{J}_1 | (z-L)^{-1} | \tilde{\alpha}_2 \tilde{J}_2 \rangle = \frac{1}{(2\pi)^n} \ \delta(\tilde{J}_1 - \tilde{J}_2) \sum_{\nu_1 = -\infty}^{+\infty} \frac{e^{i\nu_1(\tilde{\alpha}_1 - \tilde{\alpha}_2)}}{z - \nu_1 \cdot \tilde{\omega}(\tilde{J}_1)}\tag{20}$$

and inverting the Laplace transform gives

$$\langle \tilde{\alpha}_1 \tilde{J}_1 | e^{-itL} | \tilde{\alpha}_2 \tilde{J}_2 \rangle = \delta(\tilde{J}_1 - \tilde{J}_2) \sum_{\ell = -\infty}^{+\infty} \delta(\tilde{\alpha}_1 - \tilde{\alpha}_2 - \tilde{\omega}(\tilde{J}_1)t + 2\pi\ell).\tag{21}$$

From Equation (15) and Equation (20) we obtain

$$< \tilde{\alpha}_1 \tilde{J}_1 |E| \tilde{\alpha}_2 \tilde{J}_2 > = \frac{\delta(\tilde{J}_1 - \tilde{J}_2)}{(2\pi)^n} \sum_{\nu_1 = -\infty}^{+\infty} \frac{e^{i\nu_1(\tilde{\alpha}_1 - \tilde{\alpha}_2)}}{(\tilde{\omega}(\tilde{J}_1) \cdot \nu_1 = 0)}. \tag{22}$$

A time average or dynamical invariant which is 2π - periodic in $\tilde{\alpha}_1$ is

$$\phi(\tilde{\alpha}_1 \tilde{J}_1) = \int_{-\infty}^{+\infty} d\tilde{J}_2 \int_0^{2\pi} d\tilde{\alpha}_2 \, <\tilde{\alpha}_1 \tilde{J}_1 |E| \tilde{\alpha}_2 \tilde{J}_2 > f(\tilde{\alpha}_2 \tilde{J}_2). \tag{23}$$

From Equation (22), each ϕ is a function of $\tilde{J}_1$ alone if and only if the $\tilde{\omega}(\tilde{J}_1)$ are rationally independent, i.e., $\nu_1 \cdot \tilde{\omega}(\tilde{J}_1) = 0$ implies $\nu_1 = 0$.

We shall show that $a_{-1} = \lim za(z)$ as $z \to 0$ is an invertible operator in the P-subspace (and the result following Equation (17) will hold) if and only if $\psi(z) \to 0$ as $z \to 0$, provided that $\psi(z)$ has at most a pole as singularity at $z = 0$ and $a(z)$ has at most a simple pole as singularity at $z = 0$. We observe that for $a(z)$ to be of this form, it is sufficient that each eigenvalue $\lambda(\nu_1, \tilde{J}_1) = \nu_1 \cdot \tilde{\omega}(\tilde{J}_1)$ either equals zero or obeys $|\lambda| > \epsilon > 0$, where ϵ is independent of ν_1 and $\tilde{J}_1$, i.e., $\lambda = 0$ is an isolated eigenvalue.

To see this we note

$$a(z) = \int_{-\infty}^{+\infty} d\tilde{J}_1 \sum_{\nu_1.} P | \nu_1 \tilde{J}_1 > (z - \lambda(\nu_1 \tilde{J}_1))^{-1} < \nu_1 \tilde{J}_1 |P =$$

$$= \frac{1}{z} \int_{-\infty}^{+\infty} d\tilde{J}_1 \sum_{\nu_1} P | \nu_1 \tilde{J}_1 > <\nu_1 \tilde{J}_1 | P$$

$$(\lambda = 0)$$

$$- \sum_{\ell=0}^{\infty} \int_{-\infty}^{+\infty} d\tilde{J}_1 \sum_{\nu_1} \frac{P|\nu_1 \tilde{J}_1>}{\lambda} \left(\frac{z}{\lambda}\right)^\ell < \nu_1 \tilde{J}_1 | P$$

$$(|\lambda| > \epsilon > 0)$$

$$= \sum_{\ell=-1}^{\infty} a_\ell z^\ell \tag{24}$$

and a_{-1} = PEP. Thus, inserting $\psi(z) = \sum\limits_{\ell=-k}^{\infty} \psi_\ell \, z^\ell$ into

$(\psi(z) - z)\, a(z) = -P = a(z)(\psi(z)-z)$, and comparing coefficients of like powers of z (by uniqueness of the Laurent expansion), we obtain (Stey, 1974):

$$\psi_{-k}\, a_{-1} = 0 = a_{-1}\psi_{-k}$$

$$\psi_{-k}\, a_0 + \psi_{-k+1}\, a_{-1} = 0 = a_{-1}\, \psi_{-k+1} + a_0\, \psi_{-k}$$

$$\cdot$$
$$\cdot$$
$$\cdot$$

$$\psi_{-k}\, a_k + \ldots + (\psi_1 - 1)\, a_{-1} = -P = a_{-1}(\psi_1 - 1) + \ldots + a_k\, \psi_{-k}$$
$$(25)$$

Now if $a_{-1}Pg = 0$ implies $Pg = 0$, then Equations (25) yield successively $\psi_\ell = 0$ for all ℓ = -k, -k+1, ... , 0. Conversely, if all $\psi_\ell = 0$ for = -k, -k+1, ... 0, then Equation (25) gives $(\psi_1 - 1)\, a_{-1} = -P = a_1(\psi_1 - 1)$ which implies that $a_{-1}Pg = 0$ only if $Pg = 0$. But

$$\lim_{z\to i0^+} \sum_{\ell=-k}^{\infty} \psi_\ell\, z^\ell = 0 \quad \text{if and only if}$$

all $\psi_\ell = 0$ for ℓ = -k, -k+1, ... 0. Q.E.D. Thus, PEP g_1 = PEP g_2 implies $Pg_1 = Pg_2$ if and only if $\lim\limits_{z\to i0^+} \psi(z) = 0$ and condition

(16a) is equivalent to (16b) which is the theorem stated by Equation (17). To see that statement (16a) is equivalent to (16b) consider that 0 = PEPf implies 0 = (f, PEPf) = (Pf, EPf) = (EPf, EPf), since P and E are both orthogonal projectors; thus, EPf = 0, which in turn gives PEPf = 0, QED. Statement (16c) means that the only P space vector which is orthogonal to each dynamical invariant ϕ is Pf = 0. To show the equivalence of statements (16d) and (16c), let QLf = 0, thus, Lf = PLf and ELf = EPLf = 0 implies PLf = Lf = 0. Conversely, EPg = 0 implies that Pg is in $R(L)$ (assuming

$N^\perp(L) = R(L)$ here), Pg = Lf say. Thus, QLh = 0 implying Lh = 0 gives Pg = 0 since QLf = 0, QED. Statement (16d) means that no vector of the P space lies in the range of L. To show that (16e) and (16a) are equivalent consider that (16a) implies $Pf = a_{-1}(1-\psi_1)f = PEg$ with $g = P(1-\psi_1)f$ (assuming ψ_1 is bounded) for any Pf. Thus, $R(P)$ is in $R(PE)$ and vice

versa. Conversely, if Pf = PEg then EPEg = EPf, (g, EPEg) =
(PEg, PEg) = (Pf, Pf). Thus, EPf = 0 implies Pf = 0 which
is statement (16c), which in turn implies (16a) as we have seen
above. Statement (16e) means that each P space vector is the
P projection of some E space vector, i.e., of some dynamical
invariant.

Actually, the relation between the small z behavior of
$\psi(z)$ and the P projections of the time averages, i.e., Equation (25), has been used to determine the latter. For example,
in the finite dimensional case, $\psi(z)$ has a simple pole at
z = 0 and it was found that (Stey, 1974), see Appendix,

$$R(PE) = N(\psi_{-1}) \cap N(P_0\psi_0 P_0) \cap R(P) = R(PEP) = \qquad (26)$$
$$= N^{\perp}(PEP) = N^{\perp}(EP) = N^{\perp}((PE)^+),$$

where P_0 is the projector on to the $N(\psi_{-1})$.* In fact by means
of Equation (26), ergodicity was able to be characterized in
terms of ψ , for finite dimensional quantum systems (Stey,
1974). Such a system is ergodic if and only if there exists an
orthonormal basis for the n-dimensional space in which the
Hamiltonian acts such that when P is taken to project out all
diagonal matrix elements of operators in this basis, then (Stey, 1974

$$\dim[R(\psi_{-1}) \oplus R(P_0\psi_0 P_0)] = \text{maximum} = n-1. \qquad (27)$$

As a concrete example, we consider now a model Hamiltonian
for a system with one degree of freedom. With ω and V independent of α_1, J_1 we take

$$H(\alpha_1, J_1) = \omega J_1 - 2V \sin \alpha_1. \qquad (28)$$

By means of the transformation

$$\tilde{\alpha}_1 = \tilde{\alpha}(\alpha_1 J_1) = \alpha_1 ; \quad \tilde{J}_1 = \tilde{J}(\alpha_1 J_1) = J_1 - \frac{2V}{\omega} \sin \alpha_1 \qquad (29)$$

and Equation (20), we have

$$\langle \alpha_1 J_1 | (z-L)^{-1} | \alpha_2 J_2 \rangle = \frac{1}{(2\pi)^0} \delta[J_1 - J_2 - \frac{2V}{\omega}(\sin \alpha_1 - \sin \alpha_2)] \times$$

$$\sum_{\nu_1 = -\infty}^{+\infty} \frac{e^{i(\alpha_1 - \alpha_2)\cdot\omega_1}}{z - \nu_1 \cdot \omega}. \qquad (30)$$

* i.e., R(PE), the set of P-projections of <u>all</u> time averages
or dynamical invariants represented in the range of E is just
the intersection of the -subspace with the null space of ψ_{-1}
and with the null space of ψ_0 considered as an operator from
and to the null space of ψ_{-1}.

If we choose a P which projects out the reduced probability density in action:

$$\rho_0(J_1,t) = \frac{1}{2\pi} \int_0^{2\pi} d\alpha_1 \; \rho(\alpha_1,J_1,t) = P\rho , \qquad (31)$$

$$<\alpha_1 J_1 | P | \alpha_2 J_2> = \delta(J_1-J_2)/2\pi .$$

We obtain

$$2\pi<\alpha_1 J_1 | P(z-L)^{-1}P | \alpha_2 J_2> = \int_{-\infty}^{+\infty} dK \frac{e^{iK(J_1-J_2)}}{2\pi} (z-\tilde{\tilde{\psi}}(K,z))^{-1}, \qquad (32)$$

where

$$\tilde{\tilde{\psi}}(K,z) = z - \left[\frac{1}{i}(1-e^{\frac{2\pi i}{\omega}}z)^{-1} \int_0^{\frac{2\pi}{\omega}} dt \; e^{izt} J_0(\frac{4VK}{\omega} \sin(\frac{\omega t}{2})) \right]^{-1}$$

$$= z - \left[\sum_{\nu_1=-\infty}^{+\infty} \frac{J^2_{\nu_1}\left(\frac{2VK}{\omega}\right)}{z-\nu_1\omega} \right]^{-1} \qquad (33)$$

and $J\nu(x)$ is the ν-th order Bessel function of the first kind. Equation (6) then yields

$$2\pi<\alpha_1 J_1 | \psi(z) | \alpha_2 J_2> = \overline{\psi}(J_1-J_2,z) = \int_{-\infty}^{+\infty} dK \frac{\tilde{\tilde{\psi}}(K,z)}{2\pi} e^{iK(J_1-J_2)}. \qquad (34)$$

Looking at the behavior of the collision operator:

$\tilde{\tilde{\psi}}(K,z) \sim [1-J_0^{-2} (\frac{2VK}{\omega})]z \rightarrow 0$, (provided that $2VK/\omega$ is not a zero of $J_0(x)$) when $z \rightarrow i0^+$ with fixed $\omega\neq0$. On the other hand, if $\omega \rightarrow 0$,

$$\tilde{\tilde{\psi}}(K,z) = -i|2VK|^2/[(|2VK|^2-z^2)^{1/2} -iz] ,$$

Im z > 0, which gives $\tilde{\tilde{\psi}}(K,i0^+) = -2i|VK|\neq0$, $K\neq0$. This agrees with what is expected from the theorem of Equation (17). Here,

$$2\pi<\alpha_1 J_1 |PEP| \alpha_2 J_2> = \int_{-\infty}^{+\infty} dK \; J_0^2 (\frac{2VK}{\omega}) \frac{e^{iK(J_1-J_2)}}{2\pi} \qquad (35)$$

so that $Pf = f_0(J_1) = PEPg(J_1)$ is the inverse Fourier transform of J_0^2 (2VK/ω) times the Fourier transform of $Pg = g_0(J_1)$.

When $\omega \neq 0$, and $\psi(i0^+) = 0$, two different g_0 functions must yield two different f_0 functions, by the uniqueness theorem for Fourier transforms. However, after $\omega \to 0$, and $\psi(i0^+) \neq 0$, we have $zP(z-L)^{-1} P \to 0$ for $z \to 0$. Then, all g_0 functions give the same f_0. Also, when $\omega \neq 0$,

$$\phi(\alpha_1 J_1) = \int_0^{2\pi} d\alpha_2 f(\alpha_2, J_1 - \frac{2V}{\omega} (\sin \alpha_1 - \sin \alpha_2)) \frac{1}{2\pi}$$

is a general invariant, with f arbitrary. Thus, if some $g_0(J_1)$ is orthogonal to all ϕ:

$$0 = (g_0, \phi) = \int_{-\infty}^{+\infty} dk \, \tilde{g}^*(k) \, J_0^2 (2VK/\omega) \, \tilde{g}(k), \text{ with } \tilde{} \text{ tilde}^\wedge =$$

Fourier transform and $f(\alpha, J_1) = g_0(J_1)$ having been chosen. The integrand thus must equal zero, implying that $\tilde{g}(k) = 0$ almost everywhere and $g_0(J_1) = 0$. When $\omega = 0$, the only square integrable solution to $L\phi = 0$ is $\phi = 0$, to which all functions are orthogonal, which is expected by the equivalence of (16a) and (16c).

Studies of more complicated models will be presented elsewhere.

ACKNOWLEDGEMENTS.

The author wishes to thank Professor I. Prigogine and the Instituts Internationaux de Physique et de Chimie, fondes par E. Solvay, for support of this work.

REFERENCES.

Balescu, R. (1975), "Equilibrium and Nonequilibrium Statistical Mechanics", Wiley, N.Y.

Prigogine, I. (1962), "Nonequilibrium Statistical Mechanics", Interscience, N.Y.

Prigogine, I. and Resibois, P. (1961), Physica 27, p. 629.

Rae, J. (1973), "Recent Advances in Dynamical Astronomy", B.D. Tapley and V. Szebehely, (eds.), pp. 262-269.

Rae, J. (1969), Bull. cl. Sci. Acad. Roy. Belg. 11, pp. 980, 1040.

Ruelle, D. (1977), "Sensitive Dependence on Initial Condition
    and Turbulent Behavior of Dynamical Systems", N.Y. Aca-
    demy of Science Conference on Bifurcation Theory, preprint.

Stey, G.C. (1978), to be published.

Stey, G. C. (1974-1975), unpublished work, cited in Prigogine, I.
    (1975), Internat. Journ. Quant. Chem. Symp., 9, p. 443;
    Prigogine, I., Grecos, A. and George, Cl., Celestial
    Mechanics, 16, p. 489.

APPENDIX

We shall prove Equation (26). Each $a_n = Pa_nP$, $\psi_n = P\psi_nP$, and
each of these operators is assumed to be hermitian. When $k=1$,
the set of equations designated by "Equation (25)", becomes

$$KA = 0 = AK \tag{A1}$$

$$KB + \Lambda A = 0 = A\Lambda + BK \tag{A2}$$

$$KC + \Lambda B + MA = -P = AM + B\Lambda + CK \tag{A3}$$

where $A=a_{-1}$, $B=a_0$, $C=a_1$, $K=\psi_{-1}$, $\Lambda=\psi_0$, $M=\psi_1-1$. If X, Y, Z are the
orthogonal projectors on to $N(A)$, $N(K)$, $N((Y-Q)\Lambda(Y-Q))$,
respectively, then $X-Q=U$, $Y-Q=V$, $Z-(1-V)=W$, are the orthogonal
projectors on to $R(P)\cap N(A)$, $R(P)\cap N(K)$, $R(V)\cap N(V\Lambda V)$, respectively.
We shall prove that $1-X=W$. First, $AQ=0$ and, from Equations (A1),
(A2), we obtain $A(1-Y)=0$ and $0=A(Y+1-Y)\Lambda Y=AV\Lambda V$, which gives
$A(1-Z)=0$ also. Thus, $X(1-W)=1-W$ and $0=AX=AWX$. With Equation (A2),
$(1-Y)BWX=0$, which using Equation (A3) yields $-PWX=\Lambda YBWX+KCWX$.
From the definitions of V and W, we thus have $-(WX)^+WX=$
$XWV\Lambda VBWX+XWYKCWX=0+0$. Therefore, $(1-W)X=X=X^+=X(1-W)=1-W$, Q.E.D.
$N(A)=N(W)$ and the above results give $R(A)=R(W)=R(P)\cap N(K)\cap N(Y\Lambda Y)$
(where $Y\Lambda Y=V\Lambda V$ has been used also), which is the relation between
 (PEP) and the ψ Laurent series coefficients seen in Equation (26).
The relation $R(PE)= N^\perp(EP)=N^\perp(PEP)=R(PEP)$ follows from the
general result $R(F^+)=N^\perp(F)$, the hermiticity of E and of P and
$N(EP)=N(PEP)$. The latter is true since $EPf=0$ implies $PEPf=0$, and
conversely, $PEPg=0$ implies $0=(g, PEPg)=(Pg, EPg)=(EPg, EPg)$
giving $EPg=0$, Q.E.D.

PART III

STABILITY OF PLANETARY SYSTEMS

EVOLUTION OF ORBITS IN THE OUTER PART OF THE ASTEROIDAL BELT AND
IN THE KIRKWOOD GAPS AS INFLUENCED BY THE MASS EFFECTS OF SATURN
AND JUPITER

C. Froeschlé and H. Scholl

Observatoire de Nice, Astronomisches Rechen-Institut,
Heidelberg

ABSTRACT. It is known that the semi-major axes of the asteroids
between the orbits of Mars and Jupiter are not uniformly distributed. The depleted regions in the outer part and the Kirkwood
gaps in the inner part of the belt represent singularities in
the frequency distribution of the asteroidal semi-major axes.
The question whether these depleted regions and the Kirkwood gaps
are due to gravitational perturbations of Jupiter or due to cos-
mogonic effects has not yet been resolved.

In a series of numerical experiments (Scholl and Froeschlé,
1974, 1975), we have tried to depopulate the Kirkwood gaps at
the 2/1, 3/1, 5/2 and 7/3 commensurability by Jupiter's gravitational action on fictitious asteroids starting in a gap. In
no case die an asteroid leave any of the investigated gaps and
remained outside the gap. All the fictitious asteroids librated
through the gaps or mostly within the gaps with quite different
amplitudes and frequencies depending on the initial conditions.
In addition, no fictitious asteroids approached Jupiter closely
and therefore none escaped out of the gap. According to our
numerical experiments, the Kirkwood gaps cannot be depopulated
by Jupiter's perturbations.

Next the collisional hypothesis for the origin of the
Kirkwood gaps were tested. This assumes originally existing
asteroids or planetesimals in the gaps. Because of the especially strong perturbations, these objects vary their eccentricities strongly, and therefore can cover a large portion of the
asteroidal belt thus increasing the probability of a collision
with a belt asteroid. The larger the variation in eccentricity,
the higher the collision probability. In our test for the

115

collisional hypothesis, we used a planar Sun–Jupiter–Asteroid
model averaged by Schubart's (1964) method. Most of the calculated orbits support the collisional hypothesis. The few problematic cases which do not support the collisional hypothesis
are the almost circular orbits, the orbits starting at the
edges of the observed gaps and the orbits at the 7/3 commensurability. These problematic cases do not vary their eccentricities strongly and therefore cannot be explained by the collisional hypothesis.

Compared to the depletion of the Kirkwood gaps, the gravitational explanation for the depletion of the outer part of
the belt between the 2/1 commensurability and the Hilda family
seems to be even more difficult according to Lecar and Franklin's
(1973) numerical experiment. Lecar and Franklin caclulated orbits of fictitious objects over a few thousand years using the
elliptic planar Sun–Jupiter–Asteroid model. The expected mechanism which depletes that region is based on perturbations in the
semi-major axis and eccentricity of an asteroidal orbit which
results in orbits crossing Jupiter's. After a close approach to
Jupiter, the asteroid escapes from the considered region.

In Franklin and Lecar's experiment, the objects with higher
eccentric orbits, $e > 0.25$, escaped, while objects with $e < 0.25$
only a few had close encounters with Jupiter.

The existence of these problematic cases in both experiments, for the Kirkwood gaps and for the outer part of the belt,
suggests that the corresponding hypotheses which are mainly
based on gravitational effects are false. However, before looking for different hypotheses which would need more sophistication,
one must refine the gravitational models with respect to the
number of perturbing bodies and with respect to the periods
covered. A calculation including Saturn over much longer time
spans might reduce the number of problematic cases considerably.

Our calculations were based on Schubart and Stumpff's N-
Body Program (1966). Over 100,000 years, we computed the orbits
of fictitious asteroids which represent the problematic cases
for the collision hypothesis of the Kirkwood gaps and for the
ejection hypothesis of the outer belt. Jupiter and Saturn were
included as perturbing bodies.

For the Kirkwood gaps, our new numerical experiment yielded
the same negative result as before. The number of problematic
cases could not be reduced.

The depopulated region, $3.6 < a < 3.9$ AU, in the outer belt
was depleted by our experiment (Froeschle and Scholl, 1978) to a
larger extent than by Lecar and Franklin's experiment. The

average time-scale for the excitation of an orbit in order that
it crosses Jupiter's orbit is somewhat larger than the period
covered by these authors. After 2,000 years, 25% of all the
escapers were obtained in our experiment, 75% after 15,000 years
and 100% after 60,000 years. However, our experiment did not
depopulate the region 3.6 < a < 3.9 AU completely. Most of the
objects with starting eccentricities of e < 0.10 remained in
that region and had no close approach to Jupiter. In addition,
another family of objects which minimize their eccentricities if
their aphelia are precessing through Jupiter's orbital plane,
avoided a catastrophic encounter with Jupiter.

For both types of families, observed asteroids are known
in the range under consideration but they exist in a much lower
abundance than should be expected from our experiment. The
three asteroids (721) Tabora, (522) Helga, and PL-4164 show a
coupling between the precession of perihelia and the long period
in eccentricity. The eccentricity becomes smallest when the
perihelion and the aphelion lies in Jupiter's orbital plane.
That yields the largest possible distance to Jupiter when the
asteroid is at its aphelion. On the other hand, when the eccentricity reaches its maximum and the aphelion distance is largest
(thus yielding the smallest possible distance to Jupiter), the
argument of perihelion is close to 90° or 270°. Therefore, the
asteroid passes through its aphelion when it is high above or
below Jupiter's orbital plane. This mechanism prevents close
encounters with Jupiter.

We suppose that more objects of that kind might exist but
have not yet been detected as observers ordinarily detect minor
planets close to the ecliptic. Objects of the kind described
above, however, can be detected best at high ecliptic latitudes
when they pass through their perihelia. The discrepancy between
expected and observed asteroids might be reduced by observations.

For the first mentioned type of asteroids having e < 0.1,
more observations with very powerful instruments might reduce
that discrepancy. It is an interesting problem why we observe
so few objects with almost circular orbits in the region
3.6 - 3.9 AU as well as in the Kirkwood gaps, because gravitational models including collisions do not remove such objects
to a large extent. For the gap at the 2/1 resonsnce, Franklin,
et al., (1975) list objects with low eccentricities which seem
to librate in the gap. However, most of these minor planets
have very uncertain orbits. Further observations in order to
find these objects might reduce the discrepancy between observation and calculation.

Such long runs over 100,000 years use much computer time.
Therefore, several authors (e.g., Nacozy, 1976) proposed to

increase the masses of the perturbing bodies in order to shorten the computing time and in order to obtain perturbations with larger amplitudes. The intrinsic problem for such numerical experiments with larger masses consists in the equivalence of a Sun–Jupiter–Asteroid and a Sun–"Super Jupiter"–Asteroid model. It is not obvious that the latter model on a shorter time scale yields the same orbits as the first one. The perturbing mass may not be increased too greatly.

For our purposes, and especially for the depopulation of the outer belt, it is necessary to keep the Hilda family stable in an experiment with a "Super Jupiter", since the Hilda family is actually observed. In order to determine a limiting Jupiter mass up to which the Hilda family remains stable, we computed orbits around the 3/2 commensurability with different Jupiter masses.

According to our calculations, the model with a Jovian mass of 0.007 solar masses yields no stable orbits. All the objects escape. Therefore, a value of 0.005 solar masses might be used for a Super Jupiter without significantly destroying the topology of the model. We repeated the experiment for the region 3.6 < a < 3.9 AU with a value of 5 times Jupiter's mass ($\lambda = 5$) in order to compare the depletion curves for the Sun–Jupiter–Asteroid and the Sun–Super Jupiter–Asteroid model. The orbits of 47 fictitious objects were calculated. The following Table 1 shows the total number of escapers after certain time intervals given in years.

Table 1

Sun–Jupiter ($\lambda=1$)		Sun–Super Jupiter ($\lambda=5$)	
Time	Total Number of Escapers	Time	Total Number of Escapers
0	–	0	–
1,000	3	100	7
2,000	5	200	12
3,000	6	500	17
10,000	10	1,000	26
15,000	16	2,500	33
60,000	19		
100,000	19		

Obviously, the depletion of both models is different. Most of the escapers in the model with $\lambda =1$ were found after 15,000 years, while in the model with $\lambda=5$, the 2,500 years do not seem to be sufficient to obtain all the escapers. The us of a Super Jupiter destroys the protection mechanisms which in the model with $\lambda=1$ avoid close encounters with Jupiter. Therefore, the model with a Super Jupiter yields too many escapers and is therefore not equivalent to the original model $(\lambda=1)$.

ACKNOWLEDGEMENTS.

The work was partially supported by Grant No. 3705 of the ATP Planetologie (France).

REFERENCES.

Franklin, F.A., Marsden, B.G., Williams, J.G., Bardwell, C.M. (1975), Astron. J. 80, p. 729.

Froeschlé, C. and Scholl, H. (1978), Astron. and Astrophys., to be published.

Lecar, M. and Franklin, F.A. (1973), Icarus 20, p. 422.

Nacozy, P.E. (1976), Astron. J. 81, p. 787.

Scholl, H. and Froeschlé, C. (1975), Astron. and Astrophys. 42, p. 457.

Scholl, H. and Froeschlé, C. (1974), Astron. and Astrophys. 33, p. 455.

Schubart, J. and Stumpff, P. (1966), Veröff.d.Astron. Rechen-Inst., No. 18.

Schubart, J. (1964), SAO Special Report, No. 149.

ON THE STABILITY OF RESONANT MOTION IN THE LIGHT OF
THE REGULARIZING FUNCTION

B. Garfinkel

Yale University, New Haven, Connecticut, USA

ABSTRACT.

The regularizing function that removes the Poincaré singularity in a solution of a resonance problem is closely related
to the frequency of the motion and the associated characteristic
exponents. It may therefore have an intimate connection with
the question of stability.

1.   ON THE SINGULARITIES IN THE SOLUTION OF A RESONANCE PROBLEM.

Let the Hamiltonian of the problem be expanded in the form

$$F = F_o + F_1 + F_2 + \ldots$$

where $F_i = 0(\varepsilon^i)$ and ε is a small parameter. It is a fact
that the <u>classical</u> solution of the problem, expanded in powers
of ε, carries the divisor

$$D = j \cdot n$$

where j is a set of intergers in the range

$$- \infty < j < + \infty \, ,$$

and n is the set of the natural frequencies of the motion.

While there always exist integers j for which the denominator D is as small as we please, the corresponding numerator

121

Victor G. Szebehely (ed.), Instabilities in Dynamical Systems. 121-127.

is subject to the D'Alembert characteristic, and it approaches zero as $|j| \longrightarrow \infty$. Accordingly, we define <u>resonance</u> by the conditions

$$D \sim 0 \; , \quad |j| = 0 \; (1) \; .$$

Then the classical solution carries a resonant term with a critical divisor D, which is not compensated by a small numerator. The <u>classical singularity</u> 1/D at D = 0 corresponds to an exact commensurability of the set n.

If the resonance is <u>simple</u>, the critical divisor can be removed from the solution by means of the <u>Bohlin</u> <u>technique</u> of expanding the solution in powers of $\sqrt{\varepsilon}$. However, it was noted by Poincaré (1893) that while the Bohlin technique removes the classical resonance singularity, it gives rise to a new singularity of its own. The latter singularity is of the form $1/\dot{x}$, where x is the slowly-changing <u>critical argument</u>, librating in the range

$$x_1 \leq x \leq x_2 \; ,$$

with its time-derivative $\dot{x}$ vanishing at the <u>turning points</u> $x = x_1$ and $x = x_2$ of the libration.

2.   THE GENESIS OF THE POINCARÉ SINGULARITY.

That the classical singularity arises from formal integration of the equations of motion is well understood. The genesis of the Poincaré singularity will be now explained in the context of the Ideal Resonance Problem. The latter is defined by the Hamiltonian

$$F_o = B(y) + 2\mu^2 A(y) f(x) \tag{1}$$

(Paper X-1976(, where f_o is a 2π-periodic function of the critical argument x, y is the conjugate momentum, and $\varepsilon = 2\mu^2$. If the perturbations arising from F_1, F_2, ... are calculated by the Lie-Hori method (1966) with the new Hamiltonian chosen so that $F_1'=F_2' = \ldots = 0$, then the generator S_2 takes the form

$$S_2 = \{ \int [F_2 + \tfrac{1}{2} (F_1,S_1] \; dt \} \tag{2}$$

(See Paper XI-1977, p. 372) The integrand of (2) is a function of x, y, ξ, η,..., where ξ, η, ... are the additional variables of the full problem. Without any loss of generality, the libration center may be defined by $x = x_o$, $y = \xi = \eta = 0$, so

that y, ξ, η, are small quantities in the libration regime. Accordingly, the integrand of (2) can be expanded into a Taylor series about $y = \xi = \eta = 0$, of the form

$$\Psi = \epsilon^2 \psi(x) + \dots , \tag{3}$$

where the coefficients are function of x only. The leading term of (3) on substitution into (2) gives rise to the quadrature term,

$$S_2 = \{\epsilon^2 \int \psi(x)dx/\dot{x}(x) + \dots \tag{4}$$

where $\dot{x}(x)$ is the unperturbed value of $\dot{x}$, expressible as a function of x in virtue of (1). Therefore, the corresponding perturbation $\delta_2 y$ in y is given by the expression

$$\delta_2 y = S_{2x} = \epsilon^2 \psi/\dot{x} , \tag{5}$$

which carries a singularity at $\dot{x} = 0$. This is the singularity that is appropriately named after Poincaré.

3. THE REGULARIZING FUNCTION $\psi(x)$.

An obvious way to remove the Poincaré singularity in (5) is to transfer the term $\epsilon^2 \psi(x)$ of (3) to the unperturbed Hamiltonian F_o of (1). Then the latter can be put into the form

$$F_o = B(y) + 2\mu^2 A(o)f(x) ,$$

where f is defined by

$$f = f_o(x) + \epsilon\psi(x)/A(o) , \tag{6}$$

and the "overflow" terms $2\mu^2 A(o)y + \dots$ are transferred to F_1', ... It is the term $\psi(x)$ that we designate as the

regularizing function.

4.   THE PERIOD OF THE MOTION.

The period   T   in the second-order solution of the Ideal Resonance Problem can be extracted from equations (47) and (25) of Paper X-1976 in the form

$$T(\alpha, \mu) = (\mu\Omega)^{-1} \oint \frac{1 - 4\mu^2 N_2 (\alpha^2 - f_o)/B''}{\sqrt{\alpha^2 - f_o}} \, dx \, , \quad (7)$$

with

$$\Omega \equiv |4AB''|^{1/2} \, , \quad N_2 \equiv \frac{1}{24} (3b_4 - 5b_3^2), \quad b_i \equiv \frac{B^{(i)}}{B^{(2)}} \, ,$$

where the superscripts on   B   denote the order of the derivative with respect to   y   evaluated at   y=0,   and α is the "energy" parameter related to the Jacobi constant.

In the perturbed problem the period is affected by the disturbing Hamiltonian F_2 of $O(\epsilon^2)$ only through the term

$$F_2 = \mu^2 \overline{a_{11}} (x) \, y^2 + \dots \, .$$

As shown in Paper XII, now in press, (7) is then replaced by

$$T(\alpha, \mu) = (\mu\Omega)^{-1} \oint \frac{1 - \mu^2/B''[4N_2(\alpha^2 - f) + \overline{a}_{11}]}{\sqrt{\alpha^2 - f}} \, dx$$

$$(8)$$

We shall consider the special case $\alpha = 0,$ corresponding to small osciallations about the libration center $x = x_o,$ $y = 0.$ Since $f(x_o) = o,$ (8) reduces to

$$T(o, \mu) = (\mu\Omega)^{-1}[1 - \mu^2 \overline{a}_{11}(x_o)/B''] \lim_{\alpha^2 \to o} \oint \frac{dx}{\alpha^2 - f}$$

$$(9)$$

With the aid of the Taylor series,

$$f(x) = f(x_o) + \xi f'(x_o) + \frac{1}{2}\xi^2 f''(x_o) + \dots ,$$

$$\xi = x - x_o$$

with

$$f(x_o) = f'(x_o) = 0 ,$$

we deduce

$$\lim_{\alpha \to o} \oint \frac{dx}{\sqrt{\alpha^2 - f}} = 2\pi \left|\frac{1}{2}f''(x_o)\right|^{1/2} ,$$

so that (9) becomes

$$T(o,\mu) = (\mu\Omega)^{-1}[1 - \mu^2 \bar{a}_{11}(x_o)/B''] \left|\frac{1}{2}f''(x_o)\right|^{-1/2} 2\pi . \tag{10}$$

The corresponding angular frequency is therefore,

$$\omega(o,\mu) = \mu\Omega \left|\frac{1}{2}f''(x_o)\right|^{1/2} [1 + \mu^2 \bar{a}_{11}(x_o) + \dots] . \tag{11}$$

The explicit dependence of ω on the regularizing function follows from (6), which yields

$$\omega(o,\mu) = \mu\Omega \left|\frac{1}{2}f''(x_o)\right|^{1/2}\{1 + \mu^2[\psi''(x_o)/Af''(x_o) + \bar{a}_{11}(x_o)/B'']\}$$

$$\tag{12}$$

up to terms of $O(\mu^3)$.

5. A CHECK OF THE THEORY.

As a check of equation (12) consider the motion of the Trojan asteroids treated as the case of 1:1 resonance in the restricted problem of three bodies. As shown in Paper XI, the following identification of symbols holds:

$$\mu = \sqrt{m/2} \ , \quad A = 1, \quad B'' = 3 \ , \quad \Omega = \sqrt{12} \ , \quad x = \lambda \ ,$$

$$f_p = \frac{1}{2s} + 2s^2 - \frac{3}{2} \ ,$$

$$\psi = \frac{1}{96} s^{-4} (12 - 13s^2)(8s^3 - 1)^2 \ , \tag{13}$$

where

$$s = \sin \frac{\lambda}{2} \ , \qquad \lambda_o = \pi/3 \ .$$

We now calculate

$$f''(x_o) = \frac{9}{4} \ , \qquad \psi''(x_o) = \frac{315}{16} \ . \tag{14}$$

Furthermore, as shown in Paper XII,

$$\overline{a}_{11} = -6\overline{a}_{11} = -\frac{1}{2}(s^{-3} + 2s^{-1} + 16 - 40s^2) \ ,$$

which leads to

$$\overline{a}_{11}(x_o) = -9 \tag{15}$$

On substitution from (13-15), the formula (12) now becomes

$$\omega(o,m) = \sqrt{\frac{27}{4}} \ m \ (1 + \frac{23}{8} m + \dots) \ , \tag{16}$$

which is in agreement to $0(m^{3/2})$ with the classical expression

$$\omega(o,m) = \{\frac{1}{2}[1 - \sqrt{1 - 27m(1-m)}\}^{1/2} \tag{17}$$

(Szebehely, 1967) for the frequency of small oscillations about the Lagrangian point L_4.

6. CONCLUSION.

It thus appears that the regularizing function $\psi(x)$ not only removes the Poincaré singularity in a solution of a resonance problem but it also yields the value of the frequency that is correct to the order of the perturbation theory. Since the frequency is related to the characteristic exponents of the variational equations, it can be reasonably expected that the regularizing function is intimately connected with the question of the stability of the motion.

7. REFERENCES

Garfinkel, B. (1976), Cel. Mech. 13, 229 (Paper X).

Garfinkel, B. (1977), Astron. J. 82, 368 (Paper XI).

Garfinkel, B. (1978), Cel. Mech. (in press - Paper XII).

Hori, G. (1966), Publ. Astron. Soc. Japan, 18, 287.

Poincaré, H. (1893), "Les Méthodes Nouvelle de la Mécanique Céleste", Vol. 2, p. 206, Dover, New York.

Szebehely, V. (1967), "Theory of Orbits", p. 250, Academic Press, New York.

THE RINGS OF SATURN AND URANUS

Peter Goldreich
California Institute of Technology, Pasadena, California, USA

Scott Tremaine
Institute of Astronomy, Cambridge, England

1. INTRODUCTION.
We now know of two systems of planetary rings. These systems are particularly interesting for at least two reasons. First, there are striking contrasts between the two systems: the rings of Saturn are bright, broad rings separated by narrow gaps, while the rings of Uranus are dark, narrow and widely spaced. Second, the rings are the nearest disk-like systems we have, and should offer valuable insights into the dynamics of more exotic and less accessible flat systems such as spiral galaxies, accretion disks and the primordial solar system.

We will briefly review the observations and theory relevant to the dynamics of the two ring systems.

2. THE RINGS OF SATURN.

a) Observations

There are two, bright, broad rings separated by a narrow gap (the Cassini division). A third faint ring lies inside the bright rings. The optical depth of the bright rings normal to the ring plane is between 0.1 and 1 (Cook and Franklin, 1958). The optical depth at ratio wavelengths is similar (Briggs, 1974).

Observations during the earth's periodic passage through the ring plane show that the ring thickness is less than 3 km (Bobrov, 1970).

The far infrared brightness temperature of the rings is

Victor G. Szebehely (ed.), Instabilities in Dynamical Systems. 129-133.

roughly 90 K (Murphy, 1974), consistent with the equilibrium
thermometric temperature for high albedo particles at Saturn's
distance from the sun. However, their microwave brightness
temperature is $\lesssim$ 10 K (Briggs, 1974).

Radar measurements show that the rings are excellent re-
flectors at microwave wavelengths (Goldstein and Morris, 1973).

b)   Theory

The collision time t_c of the ring particles is related
to the optical depth τ and the orbital frequency Ω by
$\Omega t_c \sim \tau$. Thus, the ring particles collide every few hours. Collisions cause the ring to spread. The characteristic diffusion
time t_D is comparable to the time a particle requires to random walk across the ring (Brahic, 1977; Goldreich and Tremaine,
1978a),

$$t_D \sim t_c (\Delta r / \delta r)^2 . \tag{1}$$

Here Δr is the radial width of the ring and δr is the typical
semimajor axis difference of colliding particles. Survival of
the rings requires that t_D exceeds the age of the solar system. The ring thickness Z is roughly equal to δr. Thus
equation (1) yields $Z \lesssim$ 10 meters .

The ring particles are probably composed largely of water
ice. In this case the low radio emission implies that the mean
particle radius is $\lesssim$ 0.1 absorption lengths or perhaps $\lesssim$ 10m.
There is a composition-independent lower limit of 1 cm radius,
set by the high radar reflectivity.

For ice particles collisions are likely to be inelastic
even for impact velocities as low as 10^{-3} cm s^{-1}. This implies
that a ring of ice particles would be a monolayer (Goldreich
and Tremaine, 1978a). Thus, the two lines of evidence from the
preceding two paragraphs suggest that the ring thickness Z is
between 1 cm and 10 m.

The outer ring shows an azimuthal brightness variation
which is asymmetric with respect to the Earth-Saturn center line.
Brightness maxima and minima precede and follow conjunctions of
the ring particles with this line (Reitsema et al, 1976; Lumme
and Irvine, 1976). No convincing explanation of this phenomena
exists. Suggestions include albedo variations on synchronously
rotating particles and gravitationally enhanced, non-radial

particle density fluctuations (Colombo et al, 1976).

The inner edge of the Cassini division is near the position of the 2:1 orbital resonance with the satellite Mimas. The problem is that the gap is roughly 100 times larger than the region where Mimas is expected to have a strong effect. The resolution of this difficulty (Goldreich and Tremaine, 1978b) is based on the collective response of the ring particles. Mimas excites a trailing spiral density wave at the resonance.

As the wave is damped, it transfers negative angular momentum to the ring particles. Consequently, the particles just outside the 2:1 resonance move inward, opening a gap.

3. THE RINGS OF URANUS.

   a) Observations

The Uranus is encircled by at least 9 narrow rings. The rings were discovered during occultation observations (Elliot et al, 1977; Millis et al, 1977; Elliot et al, 1978; Persson et al, 1978), and most of our information has come from these observations. However, the rings have also been seen on Stratoscope pictures of Uranus (Sinton, 1977; Colombo, 1977) and in direct mapping of the Uranus system at 2.2 μm, where methane bands drastically reduce the brightness of the planetary disk (Matthews and Neugelbauer, 1978).

The inner 8 rings are of order 10 km wide and have eccentricities $< 10^{-3}$. The width of the outermost or ϵ ring varies linearly with radius, a fact most naturally accounted for by a small increase of eccentricity with semi-major axis. Its mean eccentricity is 7.8×10^{-3}. The positions of all ϵ ring crossings are well fit by a precessing Keplerian ellipse with $\tilde{\omega} = 1°.37$ day^{-1}, (Nicholson et al, 1978). The profile of the ϵ ring is remarkably similar at different locations, and shows sharp edges.

   b) Theory

The lifetime of the rings is probably at least comparable to the age of the solar system because diffuse rays are not seen and there is very little inter-ring material. Arguments based on Equation (1) then suggest that in the absence of confining forces the ring particle size must be < 1 cm. However, such small particles would suffer substantial orbital decay due to the Poynting-Robertson effect. Thus, the presence of confining forces seems likely.

Attempts to relate the positions of the rings to a series of orbital resonances with the Uranian satellites (Dermott and Gold, 1977) have not been successful (Aksnes, 1977; Goldreich and Nicholson, 1977). We believe that small, as yet undiscovered, satellites near the rings are a more promising source of confining forces.

Any satisfactory theory of the Uranian rings must explain this confinement against collisional diffusion and Poynting-Robertson drag. It must also explain the ability of the ε ring to precess as a unit and the sharp edges observed in the ε ring optical depth profile.

4. ACKNOWLEDGEMENTS.

This work was supported by NASA Grants NGL-05-002-003 and NGL-05-002-140 and by NSF Grant AST 76-24281.

5. REFERENCES.

Aksnes, K. (1977), Nature 269, p. 783.

Bobrov, M.S. (1970), "Surfaces and Interiors of Planets and Satellites", A. Dollfus (ed.), Academic Press, New York, p. 377.

Brahic, A. (1977), Astron. Astrophys. 54, p. 895.

Briggs, F.H. (1974), Astrophys. J. 189, p. 367.

Colombo, G. (1977), Sky Telesc. 54, p. 188.

Colombo, G., Goldreich, P. and Harris, A.W. (1976), Nature 264, p. 344.

Cook, A.A. and Franklin, F.A. (1958), Smithsonian Contrib. Astrophys. 2, p. 377.

Dermott, S.F. and Gold, T. (1977), Nature 267, p. 590.

Elliot, J.L., Dunham, E. and Mink, D. (1977), Nature 267, 328.

Elliot, J.L., Dunham, E., Wasserman, L.H., Millis, R.L. and Churms, J. (1978), submitted to Astron. J.

Goldreich, P. and Nicholson, P. (1977), Nature 269, p. 783.

Goldreich, P. and Tremaine, S. (1978a), Icarus 34, p. 227.

Goldreich, P. and Tremaine, S. (1978b), Icarus 34, p. 240.

Goldstein, R.M. and Morris, G.A. (1973), Icarus 20, p. 260.

Lumme, K. and Irvine, W.M. (1976), Astrophys. J. Lett. 204, L55.

Matthews, K. and Neugebauer, G. (1978), Private communication.

Millis, R.L., Wasserman, L.H. and Birch, P. (1977), Nature 267, p. 330.

Murphy, R.E. (1974), The Rings of Saturn, F.D. Palluconi (ed.), NASA, Washington, D.C., p. 65.

Nicholson, P., Persson, E., Matthews, K., Goldreich, P. and Neugebauer, G. (1978), submitted to Astron. J.

Persson, E., Nicholson, P., Matthews, K., Goldreich, P., and Neugebauer, G. (1978), IAUC, No. 3125.

Reitsema, H.J., Beebe, R.F., and Smith B.A. (1976), Astron. J. 81, p. 209.

Sinton, W.M. (1977), Science 198, p. 503.

INSTABILITIES IN PERIODIC PLANETARY-TYPE ORBITS

John D. Hajidemetriou

University of Thessaloniki, Thessaloniki, Greece

ABSTRACT. The purpose of this paper is to study the mechanism by which instabilities are generated in planetary systems. All orbits are considered to be planar. A planetary system is considered as a periodic motion of the general N-body problem where one or two bodies, called the Suns, have masses much larger than the masses of the other bodies, called planets. In the case of one Sun and two planets, in the direct revolution case, it is proved that for nearly circular orbits of the planets instabilities are generated only at the resonances of the form 1/3, 3/5, ... (2n-1)/(2n+1),...

In the retrograde case all planetary systems are stable. In the case of one Sun and three or more planets it is shown that there correspond two or more periodic orbits which differ in phase only. Instabilities can be generated in all cases but there stable configuration also exist. For planetary systems with two Suns and two or more planets, it is proved that instabilities can always occur, but stable configurations also exist. Several numerical examples are given.

1. INTRODUCTION.

A planetary system is a system of N bodies P_1, P_2, ...P_N, with finite masses m_1, m_2, ...m_N, which move under their mutual gravitational attraction. In this system, the mass m_2 of P_2 is considered much larger than the masses of the other bodies

135

Victor G. Szebehely (ed.), Instabilities in Dynamical Systems. 135-163.

P_1, P_3,...P_N. The body P_2 will be called the Sun and the
bodies P_1, P_3,...P_N will be called planets. The main problem
is to find the properties of the motions of each planet. Since
the motion of a planet is coupled to the motion of all the other
planets, this problem is equivalent to the solution of the general N-body problem, which is not known in general. Thus, we
must make some simplifying assumptions.

The simplest assumption, in order to find the orbit of a
particular planet, say P_i, is to ignore the effect of all the
other planets. Thus we have the well known two-body problem for
the Sun, P_2, and the planet, P_i, and the orbit of P_i is an
elliptic orbit (for bounded motion). This unperturbed orbit is
stable, and the main problem is whether the effect of the other
planets will destroy the stability.

Let us see now how one could take into account the effect
of the other planets on the motion of a particular planet, P_i.

We shall not discuss the classical planetary theory where the
orbit of each planet is considered as a Keplerian orbit whose
elements are perturbed by the atrraction of the other planets,
but we shall concentrate on planar periodic orbits of the planetary N-body system (the Sun and N-1 planets) and study their stability. Several simplifying assumptions are made.

(a). Restricted circular 3-body problem. In this approximation
we study the motion of a small planet under the gravitational
effect of the Sun and Jupiter. The bodies P_2 (Sun) and P_1
(Jupiter) revolve around their common center of mass in circular
orbits and the body P_3 (small planet), considered as a massless
body, moves under the gravitational attraction of the two primaries P_2 and P_1 (c.f. Szebehely, 1967). It is known that
several families of periodic orbits of P_3 exist in a rotating
frame of reference. If the mass of Jupiter (m_1) were equal to
zero, the orbit of P_3 would be Keplerian, in the rotating
frame. When $m_1 > 0$, the orbit of P_3 is perturbed, (the
orbits of P_1 and P_2 are not affected) and this perturbation

is taken into account in this approximation.

(b). Elliptic restricted 3-body problem. We have the same approximation as above. The only difference is that the Sun and Jupiter revolve in elliptic orbits around their common center of mass. As in the circular case, families of periodic orbits of P_3 exist in the inertial frame, along which the eccentricity of

the primaries varies (Broucke, 1969).

In the approximations (a) and (b) we can consider any number of planets (bodies P_3, P_4...) because the planets are massless and consequently they do not interact with each other, and also they do not influence the motion of the primaries. In many cases, however, the effect of the mass of the planet P_3 on the

Sun and on the other planets is not negligible. Thus, a better approximation is the

(c). general planetary 3-body problem. In this approximation the gravitational effect of all three bodies on each other is taken into account. For zero masses of P_1 and P_3, their

orbits are Keplerian around the Sun (P_2) and both are perturbed

when $m_1, m_3 > 0$. In this case families of periodic orbits of

the 3-body system also exist (for fixed mass of all the bodies), in a rotating frame (Hadjidemetriou, 1976; Delibaltas, 1977). Some of these orbits simulate actual motion in the Solar system.

(d). general planetary N-body problem (N>4). This is an extended case of the above, obtained from the former by the addition of the planets P_4, P_5... We also have families of periodic orbits

of the N-body system (for fixed masses of all the bodies), in a rotating frame of reference (Hadjidemetriou, 1977, Hadjidemetriou and Michalodimitrakis, 1978), and some of them simulate actual motions in the solar system.

The addition of more planets increases the number of degrees of freedom and this complicates the problem significantly. In some cases it is possible to make the simplifying assumption that the mass of a planet is negligible, and as a consequence its motion is uncoupled from the motion of the rest of the system. In particular we may consider the

(e). planetary restricted 4-body problem.

The underlying assumption is that the main perturbation on

a small planet comes from Jupiter and Saturn. Thus, we may consider a periodic motion of the system Sun-Jupiter-Saturn (see
Hadjidemetriou, 1976, 1978b) with their actual masses, in a rotating frame and study the motion of a massless body P_4 (small

planet) in this rotating frame. Since the motion of the system
Sun-Jupiter-Saturn is known, the motion of P_4 in the rotating

frame can be considered as a motion with two degrees of freedom.
This is much simpler than the general 4-body problem which has 5
degrees of freedom. It is shown (Hadjidemetriou, 1978b) that
several periodic orbits of P_4 exist, which are close to actual

motions in the Solar system.

     In all the cases we had planetary systems with one or more
planets, which have two Suns. Such systems, with nonzero masses
of the planets, describing periodic orbits in a rotating frame,
are known to exist (Hadjidemetriou, 1977). For zero masses of
the planets, the unperturbed orbits are not Keplerian orbits but
periodic orbits of the restricted 3-body problem with the two
Suns as primaries.

     In the following we shall describe first the families of
periodic orbits of the planetary type and then we shall discuss
their stability, both analytically and numerically. In particular, we show how a critical unperturbed configuration becomes
stable or unstable when it is perturbed. We shall also compare
the various approximations as far as stability is concerned.

2. PROPERTIES OF PERIODIC PLANETARY-TYPE ORBITS.

     We shall describe here, briefly, the properties of families
of periodic planetary-type orbits with nonzero masses of all the
planets. We shall distinguish two cases: N=3, i.e., the Sun
and two planets and N≥4, i.e., the Sun and three or more planets.
This distinction is useful because in the N=3 case we can present all the essential features of the continuation from a degenerate to a real periodic motion (with nonzero masses) which
can be easily extended to any number of bodies. Thus we are not
lost in unnecessary details. The second reason is that there is
a qualitative difference, as far as stability is concerned, between a system with two planets and a system with three or more
planets.

(a). Planetary systems with two planets and one Sun (N=3).

     A detailed description of this type of motion is given in
Hadjidemetriou (1976). We shall present here the main features.

These orbits can be obtained as the continuation, by increasing the masses, of a degenerate periodic motion such that two massless bodies, P_1 and P_3, revolve around the body P_2 (Sun) in circular orbits in the same direction in the plane. It can be seen that if we consider a rotating frame xoy with P_2 at the origin and the x-axis defined by the bodies P_2, P_1 (the positive direction being from P_2 to P_1), the motion of this degenerate system is periodic, in this rotating frame, symmetric with respect to the x axis, for any value of the radius R_3 of the body P_3. (We consider, at present, simple orbits only, i.e., the perpendicular intersection of P_3 with the x-axis takes place at the first intersection). The period is equal to

$$T = 2\pi/(1 - T_1/T_3) , \qquad (1)$$

provided we have used the normalization defined by $m_2 = 1$, $G=1$ and $\dot{\nu}_o = 1$, where $\dot{\nu}_o$ is the angular velocity of the rotating frame. In this normalization, the period of P_1 is $T_1 = 2\pi$ and its radius is $R_1 = 1$. From the above we see that we have a monoparametric family of degenerate periodic orbits of the three bodies with R_3 (or the period T_3) of the body P_3 as a parameter.

It is shown that the above periodic orbits can be continued to the general case m_1, $m_3 > 0$, with nearly circular orbits of the bodies P_1 and P_3, with the same period provided $T \neq 2\pi n$, $n = \pm 1, \pm 2, \pm 3 \ldots$, i.e., when $T_1/T_3 \neq n/(n+1)$. These orbits, for m_1, $m_3 > 0$ are symmetric periodic in a rotating frame xoy whose origin is at the center of mass of P_2, P_1 and its x-axis contains P_2 and P_1. In the case $T_1/T_3 = n/(n+1)$ the continuation theorem is not applicable and the numerical integrations have revealed that each resonant degenerate orbit $n/(n+1)$ generates two resonant branches, with nearly elliptic orbits of the bodies P_1 and P_3 such that the ratio $T_1/T_3 \approx n/(n+1)$ and the osculating eccentricity varies along each branch.

A symmetric periodic orbit of the the general 3-body problem, in the above rotating frame is represented by the initial conditions x_{10}, x_{30}, $\dot{y}_{30}$, and consequently, a monoparametric family is represented by a smooth curve in the above space of initial conditions. For illustration purposes however, it suffices to use the projection on the $x_{10}x_{30}$ plane only. Thus, we have the families for m_1, $m_3 > 0$ shown in Fig. 1. The degenerate family is represented in the above space by the straight line $x_{10} = 1$. When it is extended to m_1, $m_3 > 0$ it breaks to an infinite number of families.

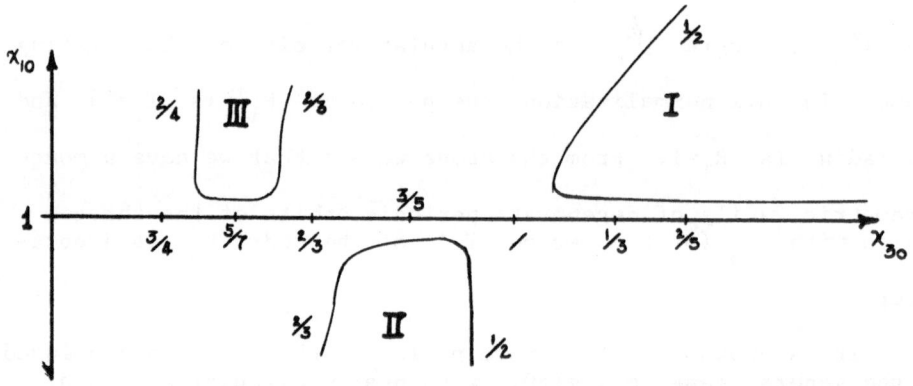

Fig. 1:   The families I, II, III of periodic planetary-type orbits (schematically). The resonances $n/(n+1)$ are indicated at each branch. The straight line $x_{10} = 1$ represents

the degenerate family along which the ratio T_1/T_3 varies. Several such values are indicated.

We note that the resonant orbits $T_1/T_3 = 1/2$, 2/3, 3/4,... separate the degenerate family $x_{10} = 1$ to an infinite number of families of periodic orbits, I, II, III,... Each family has a branch almost parallel to the straight line $x_{10}=1$, which cor-

responds to nearly circular orbits of the two bodies P_1 and P_3

and two branches (one for I) which correspond to nearly elliptic orbits of P_1 and P_3 with resonances $T_1/T_3 \approx 1/2,\ 2/3,\ 3/4$, as

indicated. These resonant branches can be considered as repre- senting families of asteroids. Also, the resonant branch 1/2 of I can be considered as representing a family of comets. There is also a periodic orbit in the "circular" part of I, which corres- ponds to $T_1/T_3 \approx 2/5$ and simulates the Sun-Jupiter-Saturn system;

The stability of these orbits will be discussed in a later sec- tion.

So far we have assumed that the circular orbits of the planets in the degenerate case, are described in the same direc- tion (in the inertial frame). If the outer planet P_3 describes

a circular orbit in a direction opposite to P_1, then the period

of the whole 3-body system in the rotating frame defined in the direct revolution case, is

$$T = 2\pi/(1 + T_1/T_3)\ ,$$

in the same normalization. Since T_1/T_3 varies between 1 and

0, the period is not a multiple of 2π, for a finite value of R_3. Thus, the retrograde degenerate family $x_{10}=1$ is extended

to the real case $m_1,\ m_2 > 0$ as a single family of periodic orbits,

symmetric with respect to the x axis.

All the above orbits are simple periodic orbits. Let us consider now, in particular, the degenerate periodic orbit $T_1/T_3 \approx 2/5$ in the direct revolution case. Its period is given

by (1) as $T=10\pi/3$. As this is not equal to a multiple of 2π, it is continued to the real case $m_1,\ m_3 > 0$, as mentioned above

(see also Fig. 1). Let us consider now this degenerate periodic orbit described three times. Its period is now equal to $T=10\pi$, i.e., a multiple of 2π, and consequently the continu- ation theorem is not applicable when $m_1,\ m_3 > 0$. The actual numeri-

cal computations have shown that this resonant, triply symmetric,

degenerate periodic orbit generates two families of triply symmetric periodic orbits of the planetary type, along which
$T_1/T_3 \approx 2/5$ and the osculating eccentricity of the two planets

varies (Hadjidemetriou, 1978b). This phenomenon is similar to
the generation of the resonant branches $n/(n+1)$ in the case
of simple orbits (see also Bozis and Hadjidemetriou, 1976).

All the above periodic orbits of the planetary type have
been obtained by the continuation of a degenerate periodic motion with <u>circular</u> orbits of the two bodies P_1 and P_3. One

could ask whether the same procedure could be used by starting
from elliptic orbits P_1 and P_3. We set the following problem:

Consider two <u>elliptic</u> orbits of P_1 and P_3 around P_2, in the

same plane, when $m_1=m_3=0$, such that $T_1/T_3=p/q$, p,q integers.

Evidently, this degenerate three-body system is periodic in an
inertial system, for <u>any</u> value of the eccentricities e_1 and e_3.
Is it possible to extend this motion to the real case m_1, m_3 0?
The answer is, in general, no as was revealed by several numerical tests. One should expect this result because otherwise we
would have families of periodic orbits of the general 3-body
problem which are two-parametric for fixed masses of all the
bodies, which is not the case (Henon, 1974; Hadjidemetriou,
1975a).

(b). <u>Planetary systems with three or more planets and one Sun</u>
        <u>(N≥4)</u>.

We shall extend now the above results to any number of
planets. We start from a degenerate system where N-1 massless
bodies P_1, P_3,...P_N revolve around the body P_2 in circular

orbits in the same direction, in the same plane. For illustration purposes we shall take N=4, but the method is easily extended to N>4. If we define a rotating frame xoy such that P_2

is at the origin and its x-axis is defined from P_2, P_1, the

above system is periodic in this frame if

$$(1-T_1/T_3)/(1-T_1/T_4) = p/q , \qquad\qquad (2)$$

where p,q are integers and T_1, T_3, T_4 are the periods of the

circular orbits of the bodies P_1, P_3, P_4, respectively, in the inertial frame. The period of this degenerate periodic motion in the rotating frame is given by

$$T = \frac{T_1}{1-T_1/T_3}\, p.$$ (3)

We shall consider only the case of symmetric periodic orbits, i.e., at $t=0$ all bodies P_1, P_3, P_4 are on the same line (the x-axis). From (2) we see that for p/q fixed we have, for each ratio T_1/T_3, a corresponding ratio T_1/T_4. Thus, for the normalization $R_1=1$, $m_1+m_2+m_3+m_4=1$, $G=1$ (also $T_1=2\pi$, $\mathbf{\nu}_0=1$), we have a monoparametric family of symmetric degenerate periodic orbits, in the rotating frame, with the radius R_3 of P_3 (or the ratio T_1/T_3) as a parameter. This degenerate family can be continued to the real case m_1, m_3, $m_4>0$ as a family of symmetric periodic orbits of the general 4-body problem with the same period, in a rotating frame xoy whose origin is at the center of mass of P_2, P_1 and the x-axis contains always these bodies (Hadjidemetriou and Michalodimitrakis, 1978). The continuation theorem is not applicable when the period, given by (3), for $T_1=2\pi$, is a multiple of 2 . The numerical integrations have revealed that in the real case m_1, m_3, $m_4>0$ the degenerate family breaks to an infinite number of families of periodic orbits, for fixed masses of all the bodies, which are "separated" by the above resonant orbits. These resonant orbits generate branches of resonant periodic orbits, as in the N=3 case for the resonances $T_1/T_3=1/2$, $2/3$, $3/4,\ldots$ (Fig. 1). As an example, we show below some resonant orbits for the degenerate family corresponding to $p/q=2/3$:

T_1/T_3	1/3	1/2	3/5	2/3	....
T_1/T_4	0	1/4	2/5	1/2	....
T	$3\times2\pi$	$4\times2\pi$	$5\times2\pi$	$6\times2\pi$	....

A symmetric periodic orbit of the general 4-body problem in the rotating frame xoy is determined by the initial conditions x_{10}, x_{30}, z_{40}, $\dot{y}_{30}$, $\dot{y}_{40}$. Thus, a monoparametric family, for fixed masses, is represented by a smooth curve in the above space of initial conditions. In this paper, for illustration purposes, we shall use the projection of the above families of planetary orbits on the plane $x_{10}x_{30}$ only. In figure 2 we show the projection of three families for m_1, m_3, $m_4 > 0$. These families are "separated" by the resonant periodic orbits mentioned above. Each family has a part which is "parallel" to the straight line $x_{10}=1$, which represents the degenerate family, and two branches (one for the family I) which correspond to resonant "elliptic" orbits with a nearly constant ratio T_1/T_3, T_1/T_4, as indicated.

We must note here that we have two different configurations, for each ratio T_1/T_3, T_1/T_4 which specifies a certain periodic orbit, which differ in phase only. These orbits may differ in their stability character, as we shall describe below. Thus, for the same masses and the same ratio p/q we have in addition to the families I, II, III,.. of Fig. 2, the corresponding families I', II', III', ... differing in phase only. We can also note that the resonant branches 1/2, 1/4 of I and II and the corresponding resonant branches of I' and II' correspond to the motion of the three inner satellites of Jupiter (Hadjidemetriou and Michalodimitrakis, 1978b).

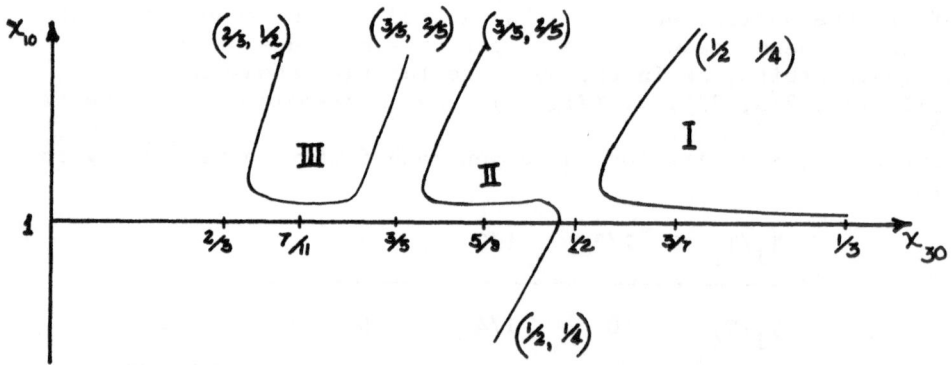

Fig. 2:   Projections of the families I, II, III of the planetary 4-body problem, for p/q=2/3 (schematically). On the degenerate family $x_{10}=1$ the resonant orbits 1/2, 3/5, 2/3 and 3/7, 5/9, 7/11 are shown.

(c). Planetary systems with two Suns and one or more planets
      (N≥3).

      It was proved in Hadjidemetriou (1975a, 1977), that we can
obtain families of periodic orbits in the general N-body problem
(N>3) such that two of the bodies have much larger masses than
the rest of the bodies, and describe nearly circular orbits.
These results were extended, for the case N=3, to elliptic motion
of the two main bodies by Christides (1978) and Kammeyer (1978).
We describe the main idea: Consider the well known restricted
circular 3-body problem, with the primaries P_1 and P_2 and N-2

massless bodies $P_3, \ldots P_N$ describing periodic orbits of the cir-

cular restricted 3-body problem, in the rotating frame, with
commensurable periods. Evidently, this degenerate system is
periodic in the rotating frame and it can be proved that, in
general, it can be continued to the real case $m_3, \ldots m_N > 0$ as a

periodic motion of the general N-body problem, in a rotating
frame. We consider planar motion only and symmetric periodic
orbits, i.e., at t=0 all the bodies lie on the x-axis. Some
numerical examples, and their stability, will be given below.

3.   STABILITY CHARACTERISTICS OF DEGENERATE PLANETARY SYSTEMS.

(a). The stability regions of the monodromy matrix.

      By the term degenerate planetary system we mean a system
of N bodies consisting of the Sun and N-1 massless bodies. We
shall discuss the case N=3 for planar motion. This is a particular case of the general planar 3-body problem which has 3 degrees
of freedom and the stability of a periodic motion is determined
from a 6x6 monodromy matrix, which however, has two unit eigenvalues, because of the energy integral. Thus, we have in fact
to study a 4x4 monodromy matrix and the corresponding 4-th order
characteristic equation (Hadjidemetriou, 1975b). This is a re-
ciprocal equation of the form

$$s^4 + a_1 s^3 + a_2 s^2 + a_1 s + 1 = 0 \qquad\qquad (4)$$

whose stability character depends on the parameters a_1 and a_2.

A detailed analysis of the characteristic equation (4) was made
by Broucke (1969). We shall present here the main features:
The equation (4) can be expressed in the form

$$(s^2 + k_1 s + 1)(s^2 + k_2 s + 1) = 0 , \qquad (5)$$

where

$$k_{1,2} = \frac{a_1 \pm \sqrt{D}}{2} , \qquad D = a_1^2 - 4a_2 + 8 . \qquad (6)$$

The parameters $k_{1,2}$ are called "stability indices" and it is evident that we have stability when $D > 0$ and $|k_{1,2}| < 2$. In this case the eigenvalues lie on the unit circle. In all other cases we have at least one eigenvalue with a real part greater than unity and consequently we have instability. The above becomes clearer if we present them in the parametric space $a_1 a_2$. It was shown by Broucke that this space can be divided into seven regions, by the straight lines $a_2 = \pm 2a_1 - 2$ and the parabola $a_1^2 - 4a_2 + 8 = 0$, and only one of them corresponds to stable motion, as described above (Fig. 3).

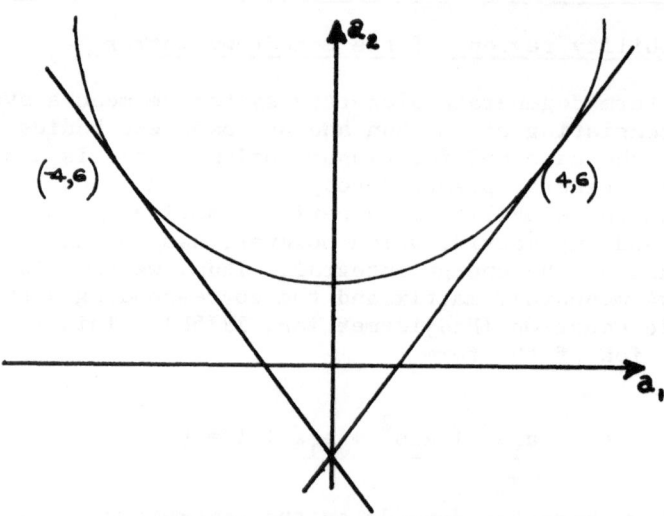

Fig. 3: The seven regions corresponding to different stability properties of the characteristic equation (4). Region 1 is stable.

We shall take now the unperturbed planetary system described in section 1 (i.e., zero masses of all the planets) and we shall find for each one the corresponding characteristic equation and its eigenvalues. In the next section we shall study their evolution when the masses of the planets are nonzero. To do so we need the stability characteristics of a keplerian orbit in a rotating frame, which are given below.

(b). Stability properties of a keplerian orbit in a rotating frame.

Let us consider the circular orbit r=a of a massless body P around P_2, with mass m_2=1, referred to a rotating frame xoy which rotates with a constant angular velocity equal to ω. The Lagrangian for the motion of P in the above rotating frame is

$$L = \frac{1}{2}\,(\dot{r}^2 + r^2(\dot{\vartheta}' + \omega)^2) + \frac{\mu}{r}\,, \qquad (7)$$

where μ=Gm_2, r is the radius from P_2 to P and ϑ is the angle between the radius vector and the rotating x-axis. The corresponding Hamiltonian is

$$H = \frac{1}{2}\,(p_r^2 + \frac{p_\vartheta^2}{r^2}\,) - \frac{\mu}{4} - p_\vartheta\omega\,, \qquad (8)$$

where

$$p_\vartheta = r^2(\omega + \dot{\vartheta})\,. \qquad (9)$$

We shall consider first the reduced problem, after the elimination of the ignorable coordinate ϑ by the use of the angular momentum integral p_ϑ=constant. The reduced Hamiltonian is (8) for p_ϑ=const, and the corresponding variational equations, corresponding to the circular motion r=a, p_r=0 will be expressed, (for reasons which will become clear later), in the form

$$i^{-1}G_2\dot{z} = H_2 z\,, \qquad (10)$$

where i= $\sqrt{-1}$, and

$$z = \begin{pmatrix} z_1 \\ z_2 \end{pmatrix} = \begin{pmatrix} \delta r \\ \delta p_r \end{pmatrix}, \quad G_2 = iJ_2 = \begin{pmatrix} 0 & -i \\ i & 0 \end{pmatrix}, \quad H_2 = \begin{pmatrix} \mu/a^3 & 0 \\ 0 & 1 \end{pmatrix} \quad (11)$$

From (10) and (11) we find that the monodromy matrix corresponding to the periodic solution $r=a$ is given by

$$Z(T) = \begin{pmatrix} \cos\omega_1 T & \frac{1}{\omega} \sin\omega_1 T \\ -\omega_1 \sin\omega_1 T & \cos\omega_1 T \end{pmatrix} \quad (12)$$

where T is the period of the circular orbit in the <u>rotating</u> frame, given by

$$T = 2\pi/(\omega_1 - \omega) , \quad (13)$$

where $\omega_1 = \mu^{1/2} a^{3/2}$ is the angular velocity of the circular orbit in the inertial frame. The eigenvalues are

$$\rho_{1,2} = e^{\pm i\omega_1 T} \quad (14)$$

and the corresponding eigenvectors are

$$g_{1,2} = (1, \pm i\omega_1)^\tau . \quad (15)$$

Note also that the stability index is

$$k_1 = -2\cos\omega_1 T. \quad (16)$$

From (13) and (14) we also see that in general $|\rho_{1,2}| \neq 1$ and $|\rho_{1,2}| = 1$ when $\omega_1 T = n\pi$. In the reduced problem, the energy integral exists but it does not have as a consequence the existence of a unit eigenvalue because it is stationary along the circular orbit.

We also find that when $\omega_1 T + (2n + 1)\pi$, we have $\rho_{1,2} = -1$ and the corresponding eigenvector is

$$g = (\lambda_1, \lambda_2)^{\tau}$$

where λ_1, λ_2 arbitrary constants.

We proceed now to the complete Hamiltonian (8), without making the reduction by the angular momentum integral. We consider again the circular orbit $r = a$ in the rotating frame xoy (angular velocity ω). The variational equations corresponding to the above periodic orbit $(r = a, p_r = 0, \vartheta = c/a^2 - \omega, p_\vartheta = c)$ can be expressed in the form

$$i^{-1} G_4 \dot{z} = H_4 z ,\tag{17}$$

where now

$$z = \begin{pmatrix} z_1 \\ z_2 \\ z_3 \\ z_4 \end{pmatrix} = \begin{pmatrix} \delta_r \\ \delta_\vartheta \\ \delta p_r \\ \delta p_\vartheta \end{pmatrix}, \quad G_4 = iJ_4 = \begin{pmatrix} 0 & 0 & -i & 0 \\ 0 & 0 & 0 & -i \\ 1 & 0 & 0 & 0 \\ 0 & 1 & 0 & 0 \end{pmatrix},$$

$$\tag{18}$$

$$H_4 = \begin{pmatrix} \omega_1^2 & 0 & 0 & -2\omega_3/\alpha \\ 0 & 0 & 0 & 0 \\ 0 & 0 & 1 & 0 \\ -2\omega_3/\alpha & 0 & 0 & 1/\alpha^2 \end{pmatrix}$$

where $\omega_3 = \mu^{1/2} a^{-3/2}$. The period of the circular orbit in the rotating frame is equal to

$$T = \frac{2\pi}{\omega_3 - \omega} ,$$

and the monodromy matrix is

(19)

$$Z(t) = \begin{pmatrix} \cos\omega_3 T & 0 & \dfrac{1}{\omega_3}\sin\omega_3 T & \dfrac{2}{a\omega_3}(1-\cos\omega_3 T) \\[2ex] -\dfrac{2}{a}\sin\omega_3 T & 1 & -\dfrac{2}{a\omega_3}(1-\cos\omega_3 T) & \dfrac{4}{a^2\omega_3}\sin\omega_3 T - \dfrac{3}{a^2}T \\[2ex] -\omega_3\sin\omega_3 T & 0 & \cos\omega_3 T & \dfrac{2}{a}\sin\omega_3 T \\[2ex] 0 & 0 & 0 & 1 \end{pmatrix}$$

From (19) we find the eigenvalues

$$\rho_{1,2} = 1, \qquad \rho_{3,4} = e^{\pm i\omega_3 T} , \tag{20}$$

or, the corresponding stability indices

$$k_1 = -2, \qquad k_2 = -2\cos\omega_3 T \tag{21}$$

The eigenvectors corresponding to $\rho_{3,4}$ for $\omega_3 T \neq \pi n$ are

$$g_{3,4} = (1, \pm 2i/a, \pm i\omega_3, 0)^\tau . \tag{22}$$

The eigenvectors corresponding to $\rho_{3,4} = -1$, for $\omega_3 T(2\pi+1)\pi$, are

$$g_{3,4} = (\lambda_1, (2/a\omega_3)\lambda_2, \lambda_2, 0)^\tau \tag{23}$$

for λ_1, λ_2 arbitrary complex constants. This implies that in this case the double eigenvalue $\rho_{3,4} = -1$ has simple elementary divisors.

   Finally we note that the eigenvalues corresponding to an elliptic orbit are all equal to unity, and the corresponding stability indices are $k_1 = k_2 = -2$. This is due to the existence of the angular momentum and the energy integrals.

(c) Stability characteristics of various degenerate planetary
     systems.

     (i). Circular restricted 3-body problem: In the zero order
approximation we have $m_1=0$ and the massless body P_3 describes
the (unperturbed) circular orbit r=a, in the same direction as
P_1. We shall refer this orbit to the rotating frame of the restricted 3-body problem, which rotates with angular velocity
$\omega=2\pi/T_1$, where T_1 is the period of the body P_1 in the inertial frame. The period T of P_3 in the rotating frame is
equal to $T=2\pi/(\omega_3-\omega_1)$, where $\omega_3=2\pi/T_3$ and T_3 is the period
of P_3 in the inertial frame. Using the theory of the previous
paragraph we find that to the circular motion of P_3 in the rotating frame there correspond the stability indices

$$k_1=-2, \quad k_2=-2\cos\left|2\pi/(1-T_1/T_3)\right| \tag{24}$$

and the corresponding eigenvalues

$$\rho_{1,2} = 1, \qquad \rho_{3,4} = e^{\pm i2\pi(1-T_1/T_3)}. \tag{25}$$

The parameters a_1, a_2 of the characteristic equation are

$$a_1=-2-2\cos(2\pi/(1-T_1/T_3)), \qquad a_2=2+4\cos(2\pi/(1-T_1/T_3)). \tag{26}$$

Thus we see that the point (a_1, a_2) lies on the line segment
AB in Fig. 3. It is at the end points A and B when
$T_1/T_3=n/(n+1)$ and $=(2n-1)/(2n+1)$, respectively. The corresponding characteristic exponents are $\rho_{3,4}=1$ and $\rho_{3,4}=-1$, respectively. In all other cases $\rho_3\neq\rho_4$. If P_3 revolves around
P_2 in a direction opposite to that of P_1, we find in a similar
way that $k_2 = -2\cos(2\pi/(1+T_1/T_3)$ and this is never equal to
±2, for $0 < T_1/T_3 < 1$, i.e., $\rho_3\neq\rho_4$. The point (a_1, a_2)

lies also on the line segment AB in Fig. 3.

(ii). Elliptic restricted 3-body problem: In the zero order approximation $(m_1=0)$ the unperturbed orbit of P_3 will be considered as elliptic in general. As mentioned above the two stability indices are given by $k_1=k_2=-2$ and the corresponding point (a_1, a_2) of the characteristic equation is the point $a_1=-4$, $a_2=6$, i.e., the point A in Fig. 3. Note that a degenerate periodic orbit of the circular restricted 3-body problem which is represented by the point A has a period equal to a multiple of $T_1=2\pi$.

(iii). Planetary 3-body problem: The planets P_1 and P_3, in the unperturbed case, $m_1=m_3=0$, describe circular orbits, in the same direction, in the same plane, around P_2. We refer the orbits of P_1 and P_3 to the rotating frame xoy whose angular velocity of rotation is $\omega=\omega_1=2\pi/T_1$. We are interested in this case in the motion of <u>both</u> P_1 and P_3, because both are perturbed when $m_1, m_3>0$. The motion of P_1 in this rotating frame is equivalent to the reduced 2-body motion along the radius vector P_2P_1 and its stability index, for the circular motion with period T, is obtained from (16):

$$k_1 = -2\cos\omega_1 T, \tag{27}$$

where

$$T = 2\pi/(\omega_1-\omega_3) = T_1/(1-T_1/T_3) \tag{28}$$

is the period of the whole 3-body system in the rotating frame. The corresponding eigenvalues are

$$\rho_{1,2}=e^{\pm i\omega_1 T}. \tag{29}$$

We have two more stability indices from the body P_3, whose motion is in this approximation uncoupled from the motion of P_1, which are, because of (20) and (21),

$$k_2 = -2, \qquad k_3 = -2\cos\omega_3 T, \tag{30}$$

and the corresponding eigenvalues are

$$\rho_{3,4}=1, \qquad \rho_{5,6}=e^{\pm i\omega_3 T}. \tag{31}$$

We note now that $\cos\omega_1 T = \cos\omega_3 T$ and consequently

$$\rho_1 = \rho_5 = e^{i2\pi/(1-T_1/T_3)} \ , \qquad \rho_{2,6} = e^{-i2\pi/(1-T_1/T_3)} \tag{32}$$

Since in the real case $m_1, m_3 > 0$ the double eigenvalue $\rho_{3,4}=1$ is conserved, because of the energy integral, we shall consider the eigenvalue $\rho_{1,2}$ and $\rho_{5,6}$ only. The corresponding characteristic equation has the stability indices k_1 and k_2 and the parameters a_1, a_2 are

$$a_1 = -4\cos\omega_1 T, \qquad a_2 = 2+4\cos^2\omega_1 T \tag{33}$$

from which we obtain

$$D = a_1^2 - 4a_2 + 8 = 0 \ . \tag{34}$$

This means that the point (a_1, a_2) lies on the part AC of the parabola in Fig. 3. Note that we are at the points A or C when $T_1/T_3 = n/(n+1)$ or $T_1/T_3 = (2n-1)/(2n+1)$ respectively.

In the retrograde revolution case, we obtain similarly

$$\rho_1 = \rho_5 = e^{i2\pi/(1+T_1/T_3)} \ , \qquad \rho_2 = \rho_6 = e^{-i2\pi/(1+T_1/T_3)}$$

(iv). Restricted planetary 4-body problem: We have here the bodies P_2, P_1, P_3 which describe a periodic motion in a rotating frame whose x-axis contains always P_2 and P_1, such

that $T_1/T_3 \cong 2/5$. The problem is to study the motion of a <u>mass-</u>

<u>less</u> body P_4 in the gravitational field of P_1, P_2, P_3. The

unperturbed $(m_1=m_3=0)$ orbit of P_4 is considered to be cir-

cular, whose period, T, in the rotating frame is given by (3)
for $T_1/T_3=2/5$. Let us consider a simple orbit of P_4, i.e.,

p=1. Then, we have from (3) $T = 5T_1/3$. The stability indices

are obtained from (21) as $k_1=-2$, $k_2=-2\cos\omega_1 T=1$, because

$\omega_1 T=10\pi/3$. The corresponding parameters a_1, a_2 are $a_1=-1$,

$a_2=0$, i.e., the point D in Fig. 3. Thus, this case seems to

be similar to the circular restricted 3-body problem, but we
shall see that there is an important qualitative difference as
far as stability is concerned.

(v). Planetary 4-body problem: This is a generalization
of case (iii). If we work in the same way we find that we have
the stability index $k_1=-2\cos\omega_1 T$ from the body P_1, and the

stability indices $k_2=-2$, $k_3=-2\cos\omega_3 T$ and $k_4=-2$, $k_5=2\cos\omega_4 T$

from the bodies P_3 and P_4 (assuming that they describe unper-

turbed circular orbits around P_2, which are referred to the

rotating frame defined in section 2b). It can be also obtained
that $k_1=k_3=k_4$. We note that we have, for this degenerate motion,

two pairs of unit eigenvalues. One pair is conserved when
m_1, m_3, $m_4>0$ because of the existence of the energy integral.

Thus we shall study the eigenvalues corresponding to k_1, k_3,

k_4 and k_5.

(vi). Planetary systems with two Suns: In the unperturbed
case $(m_3=m_4=\ldots=0)$ the orbits of P_3, $P_4,\ldots$ are uncoupled

periodic orbits of the restricted circular 3-body problem. Thus,
the orbit of each body P_3, $P_4,\ldots$ has two stability indices,

$k_1=-2$ (because of the Jacobi integral) and k_2. To these stability indices we shall also add the index $k=-2\cos\omega_1 T$ for

the motion of P_1, P_2.

   In the perturbed case m_3, m_4,...>0 one of the critical

stability indices, equal to -2, is conserved because of the energy integral. If however N>3, the rest critical stability
indices may become unstable. In the N=3 case the parameters
a_1 and a_2 of the characteristic equation behave in the same way

as in the case (i).

   In the cases (v) and (vi) the characteristic equation if
of a greater degree than 4 and a similar analysis as for the 4th
degree case is more complicated. Details of this analysis will
be given elsewhere (Hadjidemetriou and Michalodimitrakis, 1978b).

4.   EVOLUTION OF THE STABILITY WHEN THE MASSES OF THE PLANETS
     ARE FINITE.

(a). Some remarks from the theory.

     In the previous section we discussed several degenerate
planetary systems with one or two Suns and we have shown that
in the cases (i)-(iv) the stability indices lie at the boundary
of the stability region 1 (Fig. 3). The same can be shown to
hold for the cases (v) and (vi) which correspond to a higher order characteristic equation. The problem now is whether the
perturbed periodic orbit will get out of the stability region.

     The above mentioned stability indices and eigenvalues cor-
respond to the unperturbed variational equations (10) and (17),
or a combination of them (of a similar form in more dimensions).
These are Hamiltonian systems of linear differential equations
(Yakubovich and Starzhinskii, 1975, p. 104). When the masses
of the planets increase, the corresponding linear Hamiltonian
system (variational equations) is perturbed but the perturbed
system which corresponds to a new periodic orbit with non-zero
masses of the planets with the same period is, evidently, also
Hamiltonian. Also, the matrix H in (10) and (17) depends on
the periodic orbit of the original, nonlinear, Hamiltonian system
and a small change of this periodic orbit to a new periodic orbit
has as a consequence a change in the linear Hamiltonian system
(10) and (17). This is also a Hamiltonian perturbation since in
this process too, the Hamiltonian character of (10) and (17) is
conserved. From the above we see that the main problem which we
have set from the beginning, i.e., the evolution of the stability
of a degenerate periodic planetary motion, reduces to the study

of the evolution of the eigenvalues of the system (10) and (17) under a <u>Hamiltonian</u> perturbation.

The theory which we shall use has been developed by Krein and a full exposition is given in Yakubovich and Starzhinxkii (1975, ch. III). We mention briefly the main aspects of Krein's theory. This theory refers to the evolution of the eigenvalues, originally situated on the unit circle, of linear Hamiltonian systems of the form (17) under a Hamiltonian perturbation: Consider a pair of eigenvalues ρ_1, ρ_2 on the unit circle

(Fig. 4), corresponding to a Hamiltonian system (17). When the Hamiltonian system is perturbed, the pair of eigenvalues ρ_1, ρ_2

can move on the unit circle or out of it. It is shown that when $\rho_1 \neq \rho_2$ they can move only on the unit circle, otherwise the

symmetry required by the Hamiltonian nature of the system will be violated. If however we have a double eigenvalue $\rho_3 = \rho_4$

(note that in almost all unperturbed planetary orbits we have this situation) then we could have the evolution ρ_3', ρ_4' or

ρ_3'', ρ_4'', shown in Fig. 4. However Krein's theory states that the

evolution ρ_3', ρ_4', to instability, is possible only if the

eigenvalues ρ_3 and ρ_4, which coincide in the unperturbed case,

are of a different kind.

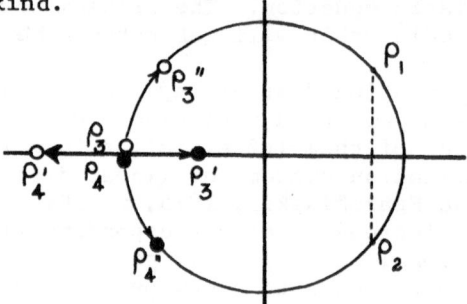

Fig. 4: The evolution of a pair of eigenvalues under a Hamiltonian perturbation.

It is first proved that the matrizant Z(t) of a Hamiltonian system (17) is G-unitary for any t, i.e., $Z_1^* = GZ^{-1}G^{-1}$.

Then an r-fold eigenvalue of the monodromy matrix Z(T) which has simple elementary divisors is classified as of the first, second, or mixed kind as follows:

If for any eigenvector, g, of the invariant subspace corresponding to the above eigenvalue we have

$$<g,g> = (Gg,g) = g*Gg>0,   (35)$$

the eigenvalue is the <u>first kind</u>. If $<g,g><0$ then it is the
<u>second kind</u> and if there exists an eigenvector such $<g,g>=0$
then it is the <u>mixed kind</u>. Thus, summarizing, we state that when
a Hamiltonian perturbation produces a meeting of eigenvalues of
the same kind on the unit circle, they cannot move off the unit
circle but if the eigenvalues are of different kinds, there
exists a Hamiltonian perturbation under which they move off the
unit circle.

(b). Application to real planetary systems.

We shall discuss the various approximations for a planetary-
orbit, mentioned in section 1.
(i). Restricted circular 3-body problem: Consider first
the direct revolution case. The unperturbed $(m_1=0)$ linear

Hamiltonian system is (17) and the corresponding monodromy matrix
is (19). We shall deal with the pair of eigenvalues

$\rho_{3,4}=e^{\pm i\omega_3 T}$ only (the pair $\rho_{1,2}=1$ is conserved when the mass

m_1 increases and/or the unperturbed orbit of P_3 varies). The

eigenvectors g_3, g_4 corresponding to ρ_3, ρ_4, respectively are

given by (22) and it can be shown (see 35) that

$$<g_3,g_3>>0,   <g_4,g_4><0 ,$$

i.e., the eigenvector corresponding to ρ_3 is of the first kind

and the eigenvector corresponding to ρ_4 is of the second kind.

When $\omega_3 T \neq (2n+1)\pi$, i.e., when $T_1/T_3 \neq (2n-1)/(2n+1)$, we have

$\rho_3 \neq \rho_4$ and consequently the perturbation will displace the $\rho_{3,4}$ on

the unit circle. This means that the stability is conserved. Con-

sider now the case $T_1/T_3 \longrightarrow (2n-1)/(2n+1)$. Then the eigenvalues

ρ_3, ρ_4 move on the unit circle and tend to meet at the point

$\rho_3 = \rho_4 = -1$. Since ρ_3 and ρ_4 are of different kinds, the unperturbed case $T_1/T_3 = (2n - 1)/(2n + 1)$ may be unstable

when $m_1 > 0$. We shall come again to this point when we shall dis-

cuss the planetary 3-body problem. In the retrograde revolution
case the eigenvalues are always distinct and consequently they
cannot go out of the unit circle.

Summarizing, we state that a circular unperturbed orbit of
a massless body cannot become unstable due to the perturbation
of another planet, with non-zero mass, which moves in a circular
direct orbit, except when the ratio of the periods of the two
planets in close to $(2n-1)/(2n+1)$, $n=1,2,\ldots$. In the retrograde revolution case the perturbation can never generate instability.

(ii). Restricted elliptic 3-body problem: We shall consider a circular orbit for the massless planet P_3 in the zero

order approximation $(m_1=0)$. The difference with the circular

case is that the monodromy matrix (19) is computed for $\omega_3 T = 2\pi n$,

because all periodic orbits are now with respect to an inertial
frame. Thus, the quadruple eigenvalue $\rho_1 = \rho_2 = \rho_3 = \rho_4$ has the

eigenvector $g = (\lambda_1, \lambda_2, \lambda_3, 0)$, with λ_1, λ_2, λ_3 arbitrary, as

can be seen from (19). Note that it does not have simple elementary divisors. This is of the mixed type and consequently
the perturbation $m_1 > 0$ or a small change of the unperturbed

periodic orbit (for example a change in phase) may make the system unstable (see Broucke, 1969).

(iii). Planetary general 3-body problem: The unperturbed
motion is the set of the two circular orbits of P_1, P_3, when

$m_1 = m_3 = 0$ in the same direction. This motion is referred to the

rotating frame mentioned before. The linear Hamiltonian system
(variational equations) is of the form (17) but it is of the
6-th order and is easily obtained from a combination of (10) and
(17). The eigenvalues are

$$\rho_1 = \rho_2 = 1, \qquad \rho_3 = \rho_5 = e^{i\omega_1 T}, \qquad \rho_4 = \rho_6 = e^{-i\omega_1 T}, \qquad (36)$$

where $\omega_1 T = 2\pi/(1 - T_1/T_3)$. The double unit eigenvalue $\rho_1 = \rho_2 = 1$

is conserved under the perturbation $m_1, m_3 > 0$ and/or a change of the unperturbed orbit of P_1, P_3 and we shall deal with the other two double pairs. We can find, for $\omega_1 T \neq n \pi$, using the eigenvectors (15) and (22) that to the pair $\rho_3 = \rho_5$ there corresponds the eigenvector $g = (\lambda_1, \lambda_2 \, 2i\lambda_2/a_3, i\lambda_1\omega_1, i\lambda_1\omega_3, 0)^T$ for the arbitrary λ_1, λ_2 and to the pair $\rho_4 = \rho_6$ the eigenvector g^*. It can be verified now that the eigenvalues ρ_3, ρ_5 are of the first kind, i.e., $<g,g>>0$, (note also that the eigenvalue a_1 in (15) and the eigenvalue a_3 of (22) are both of the first kind). Similarly, $<g^*,g^*><0$, i.e., the eigenvalues $\rho_4 = \rho_6$ are of the second kind. Thus, for $\omega_1 T \neq (2n+1)\pi$, the perturbation will not displace the eigenvalues $\rho_3 = \rho_5$ and $\rho_4 = \rho_6$ out of the unit circle, and the stability will not be destroyed. Take now the case $\omega_1 T \rightarrow (2n+1)\pi$, i.e., $T_1/T_3 \rightarrow (2n-1)/(2n+1)$. Then the double pairs $\rho_3 = \rho_5$ and $\rho_4 = \rho_6$ move on the unit circle and tend to meet at the point $\rho_3 = \rho_4 = \rho_5 = \rho_6 = -1$. Since the above two pairs of eigenvalues are of different kind, instability may occur.

Summarizing, we state that the unperturbed circular orbits of P_1 and P_3 cannot become unstable due to their mutual interaction when $m_1, m_3 > 0$ provided $T_1/T_3 \neq 1/3$, $3/5$, $5/7$.

In the latter case $T_1/T_3 = (2n-1)/(2n+1)$, the real system may become unstable, and this is the case as it was revealed by numerical computations (Hadjidemetriou, 1978b).

For the retrograde revolution case instabilities cannot arise due to the Hamiltonian perturbation. The proof is the same as in the circular restricted 3-body problem.

The above results hold for the "circular" part of the family of planetary orbits. In the "elliptic" branches we have a situation similar to the elliptic restricted problem and instabilities may occur. The numerical computations have revealed that a change in the ratio m_1/m_3 may destroy the stability.

(iv). Planetary general 4-body problem: We shall state only the difference from the above case (iii). We have, in addition to the eigenvalues (36), two more pairs, from the body P_4,

$$\rho_7=\rho_8=1, \qquad \rho_{9,10}=e^{\pm i\omega_4 T}=e^{\pm i\omega_1 T}.$$

We note that there exist now two pairs of unit eigenvalues, but only one of them is conserved when $m_1, m_3 m_4 > 0$. The other pair

does not correspond to simple elementary divisors and is of the mixed kind. Thus, the perturbation may displace it outside the unit circle, thus causing instability.

As an example, we present two periodic orbits for masses $m_2 = 1$, $m_1 = m_3 = m_4 = .001$ and the ratios $T_1/T_3 \simeq .38$, $T_1/T_4 \simeq .07$ (p/q=2/3):

x_{10}	x_{30}	x_{40}	$\dot{y}_{30}$	$\dot{y}_{40}$	stability
.99885391	−1.913	6.00002004	1.19019820	−5.59254254	stable
.99893557	1.913	5.98800716	−1.19057014	−5.57855577	unstable

These values are for the rotating frame, with the above masses normalized to $m_1+m_2+m_3+m_4=1$.

Summarizing, we state that a planetary system with more than two planets can always become unstable under a Hamiltonian perturbation.

(v). Planetary restricted 4-body problem: We have the same situation as in the case (i). Since, however, now the double unit eigenvalue is not conserved, because no energy integral exists, the perturbed orbit may become unstable, as explained in case (iv). The numerical computations have revealed that this is the case (Hadjidemetriou, 1978).

(vi). Planetary systems with two Suns: By repeating the arguments of the above cases we can show that in the case of one planet the perturbation, in general, cannot create instability. If, however, we have two (or more) planets, instability may always appear due to a perturbation (which could be a change of the mass of the planets and/or a change in phase of the unperturbed motion of the planets). As an example we study the stability of a planetary system with two Suns and two planets with masses $m_1=m_2=0.497$, $m_3=m_4=.003$. We have the following table of initial conditions in the rotating frame, for $\dot{\mathscr{v}}_o = 1$:

x_{10}	x_{30}	x_{40}	$\dot{y}_{30}$	$\dot{y}_{40}$	stability
.49911429	.36242545	−.58693744	−1.77719977	−2.30189865	stable
.49225149	.63597175	−.63597175	−1.73450657	−1.73450657	unstable
.499992	.636276	.362425	1.784738	1.766794	unstable

In all cases the planets P_3, P_4 revolve around P_1 or P_2. The above three cases differ in phase only.

6. DISCUSSION.

In this paper we studied the evolution of the stability of an unperturbed orbit under the perturbation due to the gravitational attraction of all the other planets considered, by the method of characteristic exponents. The unperturbed planetary orbit was considered to be a circular or elliptic Keplerian orbit and the application of this method was made possible because the addition of more planets to the original (unperturbed) "Sun-planet" system can result, by a suitable choice of the initial conditions, to a periodic motion of the whole system.

We have mentioned several ways by which the effect of the other planets can be taken into account, differing in the approximation made and the complexity of the problem. In the simplest approximation, the circular restricted 3-body problem, with the Sun and Jupiter as primaries, the perturbed motion of the massless planet was always stable, apart from the resonant cases 1/3, 3/5, 5/7, ... If, however, we introduce a finite eccentricity of the two primaries, the perturbed motion is allowed to become unstable. The same is true if the effect of Saturn, in addition to Jupiter, is taken into account, in the so-called planetary restricted 4-body problem. Thus, we see that the circular restricted 3-body approximation cannot reveal instabilities. On the other hand, although the restricted elliptic 3-body problem and the planetary restricted 4-body problem may drive an unperturbed (Keplerian) planetary orbit to instability, they may not yield the same result. It could happen that the same unperturbed orbit may become stable by one approximation and unstable by the other. Numerical tests must be made to study this point.

In the above three approximations we studied the motion of one planet only. This was made possible because we assumed that

it is massless and consequently it does not affect the motion
of the perturbing planets. Let us take now into account the
finite mass of all the planets. In particular, we consider the
Sun and two planets whose unperturbed (zero mass) orbits are
circular. It was found that the perturbed orbit is stable in
all cases apart from the resonances 1/3, 3/5, 5/7, ... In the
latter case a Hamiltonian perturbation exists which drives the
system to instability, so we should not expect these resonances
in a real planetary system. The same is true for the case where
the unperturbed orbits of the two planets are elliptic, so we
should not expect many elliptic orbits in planetary systems.

The above picture changes when a fourth planet is added.
In this latter case the conservation of the stability of the unperturbed orbit cannot be guaranteed, as a perturbation may make
the system unstable. This perturbation may be a change of the
masses of the planets, or a change of their mass ratio or even
a change in phase. From the above result we make the conjecture
that a planetary system with more than two planets is in general
unstable, in the sense that an arbitrary Hamiltonian perturbation may make it unstable, if the masses of all the planets are
very small. If, however, at least two of the planets are more
massive, it may happen that such a particular configuration is
stable and well inside the stability region so that an arbitrary
Hamiltonian perturbation will not be enough to drive it out to
instability.

The same distinction between N=3 and N>4 for planetary
systems with one Sun applies also to planetary systems with two
Suns. A planetary system with two Suns and one planet is, in
general, stable under a Hamiltonian perturbation but this is
not true when two, or more, planets exist.

7.   REFERENCES.

Bozis, G. and Hadjidemetriou, J.D. (1976), Cel. Mech. 13, p. 127.

Broucke, R. (1969), J. of the AIAA 7, p. 103.

Christides, T. (1978), Ph.D. Thesis, University of Thessaloniki,
        Greece.

Delibaltas, P. (1976), Astrophys. Space Sciences 45, p. 207.

Hadjidemetriou, J.D. (1978a), "Stability of Periodic Planetary-
        type Orbits of the General Planar N-Body Problem", Proceed-
        ings of I.A.U. Symposium No. 81, Tokyo, (to appear).

Hadjidemetriou, J.D. (1978b), "The Restricted Planetary 4-body Problem", Presented at the Oberwolfach Meeting, 1978, (to be published in Celestial Mechanics).

Hadjidemetriou, J.D. (1977), Cel. Mech. 16, p. 61.

Hadjidemetriou, J.D. (1976), Astrophys. Space Sciences 40, p. 201.

Hadjidemetriou, J.D. (1975a), Cel. Mech. 12, p. 155.

Hadjidemetriou, J.D. (1975b), Cel. Mech. 12, p. 255.

Hadjidemetriou, J.D. and Michalodimitrakis, M. (1978a), in: "Dynamics of Planets and Satellites and Theories of their Motion", V. Szebehely (ed.), D. Reidel Publ. Co., p. 263.

Hadjidemetriou, J.D. and Michalodimitrakis, M. (1978b), "Families of Periodic Planetary-Type Orbits in the N-body Problem and their Stability", (in preparation).

Hénon, M. (1974), Cel. Mech. 10, p. 375.

Kammeyer, P. (1978), Cel. Mech. (to appear).

Szebehely, V. (1967), "Theory of Orbits", Academic Press, N.Y.

Yakubovich, V.A. and Starzhinskii, V.M. (1975), "Linear Differential Equations with Periodic Coefficients", Vol. 1, Halsted Press.

BOUNDS ON SECULAR TERMS IN CELESTIAL MECHANICS

P. J. Message
University of Liverpool, England

1. INTRODUCTION

Studies of the equations governing the mutual gravitational
perturbations between the planets of the solar system usually
proceed, either by direct numerical integration over a delimited
period of time, or by the construction of formal series expansions. These series are in powers of the ratios of the masses of
the planets to that of the sun, for the parameters defining the
instantaneous Kepler ellipses. The series, produced by the methods developed by Laplace, Lagrange, Poisson, Poincaré, von Zeipel
and by the more recent Lie Series method, provide predictions for
the positions of the planets. The positions are verified by observation to within such limits as are properly attributable to
observational error and, also agree with the results of numerical
integrations. These latter, in the case of the four major outer
planets, have been extended to cover periods up to one million
years. The indication of these methods is that the system, considered as a system of particles moving under their mutual gravitational attractions, is stable. The major achievement of the
secular variation theory of Laplace and Lagrange was to show that,
to first order in the eccentricities and inclinations of the orbits are confined to the superposition of periodic fluctuations,
the periods being in some cases of the order of millions of years.
Poisson showed that the major semi-axes possess no secular terms
to the first or to the second order in the perturbations. We now
know that this is true to all orders, when the very long period
effects of the secular variations on the major semi-axes are
taken appropriately into account, and that the perturbation series indicate the perturbations to be comprised entirely of the
superposition of periodic fluctuations, some of very long period.

Victor G. Szebehely (ed.), Instabilities in Dynamical Systems. 165-176.
Copyright © 1979 by D. Reidel Publishing Company.

(This was developed more fully in Message, 1976.)

However, these series have been known since the time of Poincaré to be divergent in an everywhere dense set of initial conditions, due to the occurrence of denominators of arbitrarily small size, and there is no applicable mathematical criterion to determine under which conditions they may properly be used. We shall find that the error bounds provided directly from the Lagrange equations governing the elliptic elements are very unduly pessimistic, when considered in the light of the success of the perturbation series. The method of Kryloff and Bogoliuboff, used with success in the problem of the perturbed harmonic oscillator, meets, in the planetary problem, the familiar difficulty of the small divisors.

2. DIRECT DERIVATION OF BOUNDS FROM THE EQUATIONS FOR THE PERTURBATIONS.

The variation of the major semi-axis a, of the orbit of a planet P, due to the perturbation of another planet P', is governed by the equation

$$\dot{a} = (2/na) \frac{\partial R}{\partial \lambda} , \qquad (2.1)$$

where $R = \mu m' [\frac{1}{\Delta} - (r/r'^2)\cos\chi]$, the disturbing function,

   λ is the mean longitude of P,

   Δ is the distance between the planets,

   r,r' are their radial distances from the primary S,

   χ is the angle they subtend at S,

   μ is the product of the mass of S and the constant of gravitation,

   m' is the mass of P' in terms of that of S, and

$n = \mu^{1/2} a^{-3/2}$, the instantaneous mean motion of P.

If we suppose that the orbits are coplanar, we obtain

$$\dot{a} = 2nm'a [\{(a/r')^3 - (a/\Delta)^3\}\{(r'/b)\sin(\psi-\psi')$$

$$- (r'/b) e \sin(\psi'-\tilde{\omega})\}$$

$$- (a/\Delta)^3 e \sin E], \qquad (2.2)$$

where ψ and ψ' are the true longitudes of P and P', respectively, E is the eccentric anomaly, b the minor semi-axis, e the eccentricity, and $\tilde{\omega}$ the apse longitude of the orbit of P. If we suppose in particular that the orbit of P' is the larger, and that the orbits do not overlap, so that $\Delta > a'(1-e') - a(1+e)$, we have

$$|\dot{a}| < 2nm'a \left[\left\{(a/a')^3 + \left(\frac{a}{a'(1-e') - a(1+e)}\right)^3\right\} \frac{a'(1+e'+e)}{a}\right.$$

$$\left. + \frac{a^3 e}{\{a'(1-e') - a(1+e)\}^3}\right]$$

$$= m' \, f(a), \text{ say.} \tag{2.3}$$

If we restrict our consideration to periods of time during which a and a' change by no more than, say 10 percent, and e and e' remain less than, say, 0.1, and then if under these circumstances K is such that $f(a) < K$, we have

$$|a(t) - a(0)| < Km't, \tag{2.4}$$

as the crudest bound on the rate at which a can change. To proceed to a finer, second-order, bound, we need to consider the first-order perturbations in all of the orbital elements, a, λ, e, $\tilde{\omega}$, i, Ω. Suppose the Lagrange equations for each of these are written in the form

$$\dot{a} = m'f_1(a,\lambda,c_3,c_4,c_5,c_6,t)$$

$$\dot{\lambda} = n + m'f_2(a,\lambda,c_3,\ldots,c_6,t)$$

$$\dot{c}_i = m'f_i(a,\lambda,c_3,\ldots,c_6,t)$$

$$i = 3,4,5,6 \tag{2.5}$$

where $c_3 = e$, $c_4 = \tilde{\omega}$, $c_5 = i$, and $c_6 = \Omega$. Then if a_o, λ_o, c_{io}, are the initial values of a, λ, c_i, respectively, then the unperturbed expression for λ is $\lambda_u = n_o t + \lambda_o$, where $n_o = \mu^{1/2}a_o^{-3/2}$. Let us denote the first-order perturbations in a, λ, c_i by $\delta_1 a$, $\delta_1\lambda$, $\delta_1 c_i$, respectively; then we have

$$\delta_1\dot{a} = m'f_1(a_o,\lambda_u,c_{30},\ldots,c_{60},t),$$

$$\delta_1\dot{\lambda} = (dn/da)_{a=a_o} \delta_i a + m'f_2(a_o,\lambda_u,c_{30},\ldots,c_{60},t),$$

$$\delta_1\dot{c}_i = m'f_i(a_o,\lambda_u,c_{30},\ldots,c_{60},t),$$

and so

$$\delta_1 a = m' \int_0^t f_1(a_0, \lambda_u, c_{30}, \ldots, c_{60}, t)dt,$$

$$\delta_1 \lambda = \frac{-3n_0}{2a_0} \int_0^t \delta_1 a \; dt + m' \int_0^t f_2(a_0, \lambda_u, c_{30}, \ldots, c_{60}, t)dt,$$

and $$\delta_1 c_i = m' \int_0^t f_i(a_0, \lambda_u, c_{30}, \ldots, c_{60}, t)dt. \tag{2.6}$$

The crude first-order bounds are given by

$$|a(t) - a_0| < m'K_1 t,$$

$$|\lambda(t) - \lambda_u(t)| < \int_0^t n - n_0 \; dt + m'K_2 t,$$

$$|c_i(t) - c_{i0}| < m'K_i t, \tag{2.7}$$

where $|f_i| < K_i \, (i=1,\ldots,6)$

are the overall bounds of the right-hand sides of the Lagrange equations in the domain we are considering of the orbital elements. We have, by the mean-value theorem,

$$|n - n_0| < L \, |a - a_0| < Lm'K_1 t$$

where L is the greatest value of $|\frac{dn}{da}| = \frac{3n}{2a}$ in this domain, so that

$$|\lambda(t) - \lambda_u(t)| < \frac{1}{2} m'LK_1 t^2 + m'K_2 t. \tag{2.8}$$

Thus, proceeding to find a closer bound for the major semi-axis a,

$$|a(t) - \{a_0 + \delta_1 a(t)\}| = m' \int_0^t \{f_1(a, \lambda, c_3, \ldots, c_6, t)$$

$$- f_1(a_0, \lambda_u, c_{30}, \ldots, c_{60}, t)\}dt|$$

The quantity on the right-hand side of the above equation is bounded as follows:

$$\leq m' \int_0^t |f_1(a,\lambda,c_3,\ldots,c_6,t)$$

$$-f_1(a_o,\lambda_u,c_{30},\ldots,c_{60},t)|dt$$

$$\leq m' \int_0^t |M_1 m' K_1 t$$

$$+M_2(\frac{1}{2} m'LK_1 t^2 + m'K_2 t)$$

$$+ \sum_{i=3}^{6} M_i m' K_i t |dt$$

$$= m'^2 \{\frac{1}{6} M_2 LK_1 t^3 + Nt^2\} \qquad (2.9)$$

where we have used the mean-value theorem, with the bounds

$$|\frac{\partial f_1}{\partial a}| \leq M_1, |\frac{\partial f_1}{\partial \lambda}| \leq M_2, |\frac{\partial f_1}{\partial c_i}| \leq M_i,$$

and with

$$N = \frac{1}{2} \sum_{i=1}^{6} M_i K_i,$$

It is clear that the higher power of m' is offset, for very long periods of time, by the appearance of higher powers of t, and that this feature will be more marked if we proceed to higher-order perturbations by this method. Substitution of numerical values very quickly shows that these bounds are much too pessimistic, as shown by the success of general planetary theories using expressions for the perturbations of the first two or three orders, so there is need for much closer bounds. The argument of this section has been developed, with some adjustment, from Pollard (1966).

3.   METHOD OF KRYLOFF AND BOGOLIUBOFF (1943).

These authors studied the application of the variation of parameters method to a system of perturbed harmonic oscillators, governed by the equation

$$\ddot{x}_i + \omega_i^2 x_i = \epsilon f_i(x_1,x_2,\ldots,x_n) , \quad (i=1,2,\ldots,n) \qquad (3.1)$$

in which ε is to be regarded as a small parameter, and each ω_i is a constant. (Systems of equations of this type possess some of the important features of those governing planetary perturbations.) Time-dependent parameters a_i and θ_i are defined so that

$$x_i = a_i \sin\theta_i,$$

$$\text{and} \quad \dot{x}_i = a_i \omega_i \cos\theta_i, \text{ for each } i, \tag{3.2}$$

and we readily find the equations

$$\dot{a}_i = (f_i/\omega_i)\cos\theta_i,$$

$$\text{and} \quad \dot{\theta}_i = \omega_i - [f_i/(a_i\omega_i)]\sin\theta_i, \tag{3.3}$$

which govern them.

In the special case, with a single dependent variable, x,

$$f(x) = \omega^2 x^3 \tag{3.4}$$

we find

$$\dot{a} = \frac{1}{8} \omega\varepsilon a^3 (2 \sin2\theta - \sin4\theta)$$

$$\text{and} \quad \dot{\theta} = \omega - \frac{1}{8} \omega\varepsilon a^2 (3 - 4\cos2\theta + \cos4\theta). \tag{3.5}$$

From this we can derive the first-order bound

$$\left| a^2 - a_o^2 \right| < \frac{3}{2} \omega\varepsilon a_o^4 t,$$

certainly for so long as

$$t < 2/(3\omega\varepsilon a_o^2), \tag{3.6}$$

where a_o is the value of a at $t=0$.

If we put $\theta_u = \theta_o + \omega t$, θ_o being the initial value of θ, we also find

$$\left| \theta - \theta_u \right| < 2\omega\varepsilon a_o^2 t.$$

If we put

$$a_1 = a_o - \frac{1}{32} \varepsilon a_o^2 (4 \cos2\theta_u - \cos4\theta_u),$$

we may obtain the second-order bound

$$\left| a - a_1 \right| < \frac{59}{8} \omega^2 a_o^5 \varepsilon^2 t^2,$$

for at least as long as (3.6) remains satisfied. (We note the appearance of t^2, rather than t^3, since the frequency ω does not depend on a.)

We now consider the right-hand sides of (3.3) in the form of multiple Fourier series with the θ_i as arguments, and consider especially those cases in which the equation for each of the $\dot{a}_i$ has no constant term (this will certainly be so whenever f_i is an odd function of x_i). Let $\bar{B}_i(a)$ represent the constant term in the Fourier series for $\dot{\theta}_i$, and $\tilde{A}_i(a,\theta)$ and $\tilde{B}_i(a,\theta)$ the non-constant parts of the series (where θ here denotes the set of n angles θ_1, $\theta_2, \ldots, \theta_n$, and a denotes $(a_1, a_2, \ldots, a_n)$ so that the equations (3.3) may be written

$$\dot{a}_i = \epsilon \tilde{A}_i(a,\theta)$$

and
$$\dot{\theta}_i = \omega_i + \epsilon \bar{B}_i(a) + \epsilon \tilde{B}_i a, \theta), \quad (i=1,2,\ldots,n). \qquad (3.6)$$

We then define functions $u_i(a,\theta)$ and $v_i(a,\theta)$ to satisfy

$$\sum_{j=1}^{n} \omega_j \frac{\partial u_i}{\partial \theta_j} = \tilde{A}_i, \text{ and } \sum_{j=1}^{n} \omega_j \frac{\partial v_i}{\partial \theta_j} = \tilde{B}_i, \quad (i=1,2,\ldots,n)$$

$$(3.7)$$

the constants of integration being chosen so that each of the u_i and v_i has itself no constant term in its Fourier series in the θ_i. (Thus u_i and v_i correspond to the "first-order short period perturbations in a_i and θ_i, respectively.) Then a change of variables, from the a_i and θ_i to new variables $\tilde{a}_i$ and $\tilde{\theta}_i$, is defined by the equations

$$a_i = \tilde{a}_i + u_i(\tilde{a}_i, \tilde{\theta}_i)$$
$$\theta_i = \tilde{\theta}_i + v_i(\tilde{a}_i, \tilde{\theta}_i), \quad (i=1,2,\ldots,n) \qquad (3.8)$$

so that equations (3.6) become

$$\dot{\tilde{a}}_i + \sum_{j=1}^{n} \left(\frac{\partial u_i}{\partial \tilde{a}_j} \dot{\tilde{a}}_j + \frac{\partial u_i}{\partial \tilde{\theta}_j} \dot{\tilde{\theta}}_j \right) = \epsilon \tilde{A}_i \{\tilde{a} + u(\tilde{a}, \tilde{\theta}), \tilde{\theta} + v(\tilde{a}, \tilde{\theta})\}, \text{ and}$$

$$\dot{\tilde{\theta}}_i + \sum_{j=1}^{n} \left(\frac{\partial v_i}{\partial \tilde{a}_j} \dot{\tilde{a}}_j + \frac{\partial v_i}{\partial \tilde{\theta}_j} \dot{\tilde{\theta}}_j \right) = \omega_i + \epsilon \bar{B}_i \{\tilde{a} + u(\tilde{a}, \tilde{\theta})\}$$

$$+ \epsilon \tilde{B}_i \{\tilde{a} + u(\tilde{a}, \tilde{\theta}), \tilde{\theta} + v(\tilde{a}, \tilde{\theta})\},$$

$$(i = 1, 2, \ldots, n),$$

which, when use is made of equations (3.7), take the form

$$\overset{\approx}{a}_i = \varepsilon^2 \phi_i (\tilde{a}, \tilde{\theta})$$

and
$$\overset{\cdot\cdot}{\theta}_i = \omega_i + \varepsilon \bar{B}_i (\tilde{a}) + \varepsilon^2 \psi_i (\tilde{a}, \tilde{\theta}) , \quad (i=1,2,\ldots,n). \qquad (3.9)$$

Then, if we can find bounds $|\phi_i| < K_i$ we have the second-order bounds

$$|\tilde{a}_i(t) - \tilde{a}_i(o)| < \varepsilon^2 K_i t \quad (i=1,2,\ldots,n), \qquad (3.10)$$

in which only the first power of the time appears. In principle this method may be applied to each successive order in ε, giving each time a bound linear in the time. (In the special case of (3.4), we find
$$|\tilde{a}(t) - \tilde{a}(o)| < \frac{153}{64} \varepsilon^2 \omega^2 \tilde{a}(o)^5 t, \text{ and we note that}$$
$|a(t) - \tilde{a}(t)| < \frac{5}{16} \varepsilon \omega a(o)^3 .)$

The solution of the equations (3.7) to determine the u_i and v_i works well if the frequencies ω_i are all equal, or if they are exactly commensurable, so that they have a common factor, since then each term of the Fourier series for $\tilde{A}_i$ and $\tilde{B}_i$ is, on integration, simply divided by a multiple of this common value, and we may expect to find bounds for the functions u_i and v_i, and hence for the differences $|a_i - \tilde{a}_i|$ and $|\theta_i - \tilde{\theta}_i|$. If the ω_i are not commensurable, however, then the divisors on integration will be the linear combinations of them, amongst which arbitrarily small values may be found, so that no bounds are available for the u_i and v_i. This will certainly happen in any application to the problem of planetary perturbations.

4. RELATION TO LIE SERIES TRANSFORMATIONS.

The powerful method of Lie series transformation, which in many cases enables perturbation series to be derived explicitly, can be approached by considering the Taylor series giving the change over a given time interval $t_1 \leq t \leq t_2$ of any function f of a set of canonical variables $q_1, q_2, \ldots q_n, p_1, p_2, \ldots p_n$ defining a dynamical system

$$f\{q(t_2), p(t_2)\} = f\{q(t_1), p(t_1)\} + (t_2 - t_1) \left(\frac{df}{dt}\right)_{q(t_1), p(t_1)}$$

$$+ \frac{1}{2} (t_1 - t_1)^2 \left(\frac{d^2 f}{dt^2}\right)_{q(t_1), p(t_1)},$$

+ ...   (eq. 4.1)

where $q(t)$ denotes $q_1(t), q_2(t), \ldots, q_n(t)$, $p(t)$ denotes $p_1(t), \ldots,$ $p_n(t)$, and $(\phi)_{q(t_1), p(t_1)}$ indicates that the function ϕ is evaluated with $q = q(t_1)$ and $p = p(t_1)$. In terms of the Hamiltonian function H, we have

$$\frac{df}{dt} = \sum_{i=1}^{n} (\frac{\partial f}{\partial q_i} \dot{q}_i + \frac{\partial f}{\partial p_i} \dot{p}_i)$$

$$= \sum_{i=1}^{n} (\frac{\partial f}{\partial q_i} \frac{\partial H}{\partial p_i} - \frac{\partial f}{\partial p_i} \frac{\partial H}{\partial q_i})$$

$$= \{f, H\} , \qquad (4.2)$$

the Poisson bracket expression, and so in turn

$$\frac{d^2 f}{dt^2} = \{\{f, H\}, H\}$$

and thus

$$f\{q(t_2), p(t_2)\} = f\{q(t_1), p(t_1)\} + (t_2 - t_1)(\{f, H\})_{q(t_1), p(t_1)}$$

$$+ \frac{1}{2} (t_2 - t_1)^2 (\{\{f, H\}, H\})_{q(t_1), p(t_1)} + \ldots \qquad (4.3)$$

In particular,

$$q_1(t_2) = q_1(t_1) + (t_2 - t_1)(\frac{\partial H}{\partial p_i})_{q(t_1), p(t_1)}$$

$$+ \frac{1}{2} (t_2 - t_1)^2 (\{\frac{\partial H}{\partial p_i}, H\})_{q(t_1), p(t_1)} + \ldots$$

and

$$p_1(t_2) = p_1(t_1) - (t_2 - t_1)(\frac{\partial H}{\partial q_i})_{q(t_1), p(t_1)}$$

$$- \frac{1}{2} (t_2 - t_1)^2 (\{\frac{\partial H}{\partial q_i}, H\})_{q(t_1), p(t_1)} + \ldots \qquad (4.4)$$

The method depends on the fact that the relation between the canonical variables of a dynamical system at any instant are related to those at any other instant by a canonical transformation, and so any transformation from a canonical set $q_1, \ldots, q_n$, $p_1, \ldots,$ p_n to another set $Q_1, \ldots, Q_n$, $P_1, \ldots, P_n$, in which the relation between the sets can be exhibited in the same mathematical form as the relation between the $q_i(t_1)$, $p_i(t_1)$ and the $q_i(t_2)$, $p_i(t_2)$, is in fact a canonical transformation. So, if we take an appropriate function $S(q,p)$, chosen to accomplish our objectives in the transformation, and arbitrarily take $t_2-t_1 = 1$, we may replace H with S in the equations (4.4), and thus ensure that the equations

and

$$q_i = Q_i + \frac{\partial S}{\partial P_i} + \frac{1}{2} \{\frac{\partial S}{\partial P_i}, S\} + \frac{1}{3!} \{\{\frac{\partial S}{\partial P_i}, S\}, S\} + \ldots$$

(4.5)

$$p_i = P_i - \frac{\partial S}{\partial Q_i} + \frac{1}{2} \{\frac{\partial S}{\partial Q_i}, S\} - \frac{1}{3!} \{\{\frac{\partial S}{\partial Q_i}, S\}, S\} + \ldots$$

define a canonical transformation $(q,p) \rightarrow (Q,P)$.

For any function f of the canonical variables, (4.3) ensures that

$$f(q,p) = f(Q,P) + (\{f,S\})_{Q,P} + \frac{1}{2}(\{\{f,S\},S\})_{Q,P} + \ldots \quad (4.6)$$

If $H'(Q,P)$ is the Hamiltonian function for the set (Q,P), since the transformation is time-independent, we have

$$H'(Q,P) \equiv H(q,p),$$

and so (4.6) gives

$$H'(Q,P) = H(Q,P) + \{H,S\}_{Q,P} + \frac{1}{2}\{\{H,S\},S\}_{Q,P} + \ldots \quad (4.7)$$

The choice of the function S can usually be made to make H' simpler than H, if not to make some or all of the Q_i ignorable, then to leave H' free of at least a substantial part of the periodic terms. Thus S is usually of the same order of magnitude as the perturbations.

When we seek bounds, however, we may be able to make use of the fact that Taylor's theorem may be written with a remainder term:

$$f(b) = f(a) + (b-a)f'(a) + \frac{1}{2}(b-a)^2 f''(a) + \ldots$$

$$+ \frac{1}{(n-1)!}(b-a)^{n-1} f^{(n-1)}(a)$$

$$+ \frac{1}{n!}(b-a)^n f^{(n)}(\xi).$$

where ξ is some number in $a < \xi < b$. Applying this in turn to (4.3) and (4.6), we obtain an equation in which only a finite number of the terms on the right-hand side appear, the last one being evaluated with $q=\eta$, $p=\xi$, corresponding to some (unknown) point on the path from (Q,P) to (q,p) defined by the fictitious dynamical system with S as its Hamiltonian function. If then an upper bound can be found for the corresponding Poisson bracket, this is also a bound for the error in trucating the transformation series (4.6) at the appropriate finite number of terms. Thus, we have in place of (4.5), for each i,

$$q_i = Q_i + \left(\frac{\partial S}{\partial P_i}\right)_{Q,P} + \frac{1}{2}\left\{\frac{\partial S}{\partial P_i},S\right\}_{Q,P} + \frac{1}{3!}\left\{\left\{\frac{\partial S}{\partial P_i},S\right\},S\right\}_{Q,P} + \cdots$$

$$\cdots + \frac{1}{(n-1)!}\underbrace{\left\{\left\{\cdots\left\{\frac{S}{P_i},S\right\},S\right\},\ldots,S\right\}}_{\text{(n-2 brackets)}}{}_{Q,P} +$$

$$\frac{1}{n!}\underbrace{\left\{\left\{\cdots\left\{\frac{S}{P_i},S\right\},S\right\},\ldots S\right\}}_{\text{(n-1 brackets)}}{}_{\eta,\xi},$$

and

$$P_i = P_i - \left(\frac{\partial S}{\partial Q_i}\right)_{Q,P} - \frac{1}{2}\left\{\frac{\partial S}{\partial Q_i},S\right\}_{Q,P} - \frac{1}{3!}\left\{\left\{\frac{\partial S}{\partial Q_i},S\right\},S\right\}_{Q,P} - \cdots$$

$$\cdots - \frac{1}{(n-1)!}\underbrace{\left\{\left\{\cdots\left\{\frac{\partial S}{\partial Q_i},S\right\},S\right\},\ldots,S\right\}}_{\text{(n-2 brackets)}}{}_{Q,P}$$

$$- \frac{1}{n!}\underbrace{\left\{\left\{\cdots\left\{\frac{\partial S}{\partial Q_i},S\right\},S\right\},\ldots,S\right\}}_{\text{(n-1 brackets)}}{}_{\eta,\xi}. \qquad (4.8)$$

In particular, with n=1,

$$q_i = Q_i + \left(\frac{\partial S}{\partial P_i}\right)_{\eta,\xi}, \text{ and } P_i = P_i - \left(\frac{\partial S}{\partial Q_i}\right)_{\eta,\xi},$$

so that if we can find bounds $\left|\frac{\partial S}{\partial P_i}\right| \leq M_i$ and $\left|\frac{\partial S}{\partial Q_i}\right| \leq N_i$ for the region in phase space under consideration, we may conclude that

$$|q_i - Q_i| \leq M_i, \text{ and } |P_i - P_i| \leq N_i, \quad (i=1,2,\ldots,n). \qquad (4.9)$$

The perturbed harmonic oscillator problem (3.1) may be put into Hamiltonian form, provided we can find a function $F(x_1, x_2, \ldots x_n)$ such that

$$f_i = \frac{\partial F}{\partial x_i} \quad (i=1,2,\ldots,n), \qquad (4.10)$$

by writing

$$x_i = (2p_i/\omega_i)^{\frac{1}{2}} \sin q_i$$

and

$$\dot{x}_i = (2p_i \omega_i)^{\frac{1}{2}} \cos q_i \quad (i=1,2,\ldots,n), \qquad (4.11)$$

the Hamiltonian function being found to be

$$H(q,p) = \sum_{i=1}^{n} \omega_i p_i - \varepsilon \, F\{ (2p_1/\omega_1)^{\frac{1}{2}} \sin q_1, \; (2p_2/\omega_2)^{\frac{1}{2}} \sin q_2, \ldots,$$

$$(2p_n/\omega_n)^{\frac{1}{2}} \sin q_n \}. \qquad (4.12)$$

The above method may be used, though the difficulty mentioned at the end of section three prevents us from finding the bounds (4.9) if the ω_i are not commensurable.

5. REFERENCES

Krylov, N.m. and N. Bogoliuboff (1943), "Introduction to Nonlinear Mechanics." Translated by S. Lefschetz, Princeton University Press, Princeton, New Jersey.

Message, P.J. (1976) in "Long Time Predictions in Dynamics," V. Szebehely and B. Tapley (eds), Reidel Publ. Co., Dordrecht, The Netherlands, pp 279-293.

Pollard, H. (1966), "Mathematical Introduction to Celestial Mechanics," Prentice-Hall, Inc., Englewood Cliffs, New Jersey, pp 92-93.

EMPIRICAL STABILITY CRITERIA IN THE MANY-BODY PROBLEM

Archie E. Roy
Glasgow University, Scotland

ABSTRACT: In these lectures the equations of motion of the general four-body problem are derived in Jacobian coordinates.   A
reduction is made to the general three-body problem and it is
shown that, where the system consists of a binary about whose
centre of mass the third mass revolves, the equations of motion
enable two quantities ε_{12} and ε_3 to be defined as a measure of
the degree of perturbation the system of binary and third mass
suffers.   An extension is made to the four-body problem showing
that similar quantities exist.   It is conjectured that if the sum
of the magnitudes of the quantities in each equation of the system is less than a value $\ell \ll 1$, the system will be stable in the
sense that no break-up of the binary can take place for an astronomically long time.

The critical stability criterion of Zare is discussed and
the relations of ε_{12} and ε_3 to this criterion established.   Work
by Roy, Emslie and Walker, as yet unpublished, is described where
the surface of critical stability in $\varepsilon_{12}, \varepsilon_3$, α space is computed,
the quantity α being the ratio of the binary semimajor axis to
that of the third mass orbit about the binary centre of mass.
Existing data from the Solar System, known triple star systems
and numerical experiments in triple mass experiments are used to
plot all known triple systems in the $\varepsilon_{12}, \varepsilon_3$ plane, together with
the α_c contours, where α_c is the critical value of α according to
Zare's critical stability criterion.   It is found that the systems
known to be stable occupy the region $\varepsilon_3 < 10^{-3}$, $\varepsilon_{12} < 10^{-3}$, their
α values being less than the values α_c; thus providing a test for
the stability of any triple system, even if they are subsets of
n-body systems, $n > 3$.

Victor G. Szebehely (ed.), Instabilities in Dynamical Systems. 177-210.
Copyright © 1979 by D. Reidel Publishing Company.

Finally, the relevance of the existence of critical arguments in the systems Saturn–Titan–Hyperion and Sun–Neptune–Pluto to their seemingly unstable positions in the ε_{12}, ε_3, α_c plots is discussed. The plots' predictions with respect to the possible existence of hitherto undetected stable triple stellar systems and planetary systems of double stars is also discussed.

1.   THE JACOBIAN COORDINATE SYSTEM

In this section we follow Plummer's exposition. Take n point-masses $P_1, P_2, \ldots, P_n$ of masses $m_1, m_2, \ldots, m_n$ respectively (Figure 1). Let the radius vector of P_i be $\underline{r}_i$ measured from a fixed point 0 such that

$$\underline{r} = \underline{i}\, \xi_i + \underline{j}\, \eta_i + \underline{k}\, \zeta_i,$$

where $\underline{i}$, $\underline{j}$, $\underline{k}$ are unit vectors parallel to a fixed set of rectangular axes 0ξ, 0η, 0ζ through the reference point 0. Let C_i be the centre of mass of the subsystem $m_1, m_2, \ldots, m_i$ and let its radius vector with respect to 0 be $\underline{R}_i$, such that

$$\underline{R}_i = \underline{i}\, X_i + \underline{j}\, Y_i + \underline{k}\, Z_i.$$

Define $\mu_i = \sum_{j=1}^{i} m_j$,

so that

$$m_i = \mu_i - \mu_{i-1}\, ;\ \mu_o = 0.$$

$$\left.\rule{0pt}{60pt}\right\}\quad (1)$$

Then, by definition

$$\mu_i\, \underline{R}_i = u_1 \underline{r}_1 + (\mu_2 - \mu_1)\, \underline{r}_2 + \ldots + (\mu_i - \mu_{i-1})\underline{r}_i.$$

Also

$$\mu_{i-1}\underline{R}_{i-1} = \mu_1\underline{r}_1 + \ldots + (\mu_{i-1} - \mu_{i-2})\underline{r}_{i-1},$$

so that

$$\mu_i\underline{R}_i - \mu_{i-1}\underline{R}_{i-1} = (\mu_i - \mu_{i-1})\underline{r}_i. \tag{2}$$

In addition,

$$\underline{r}_1 = \underline{R}_1.$$

Now let $\underline{\rho}_i$ be the radius vector drawn from C_{i-1} to P_i.

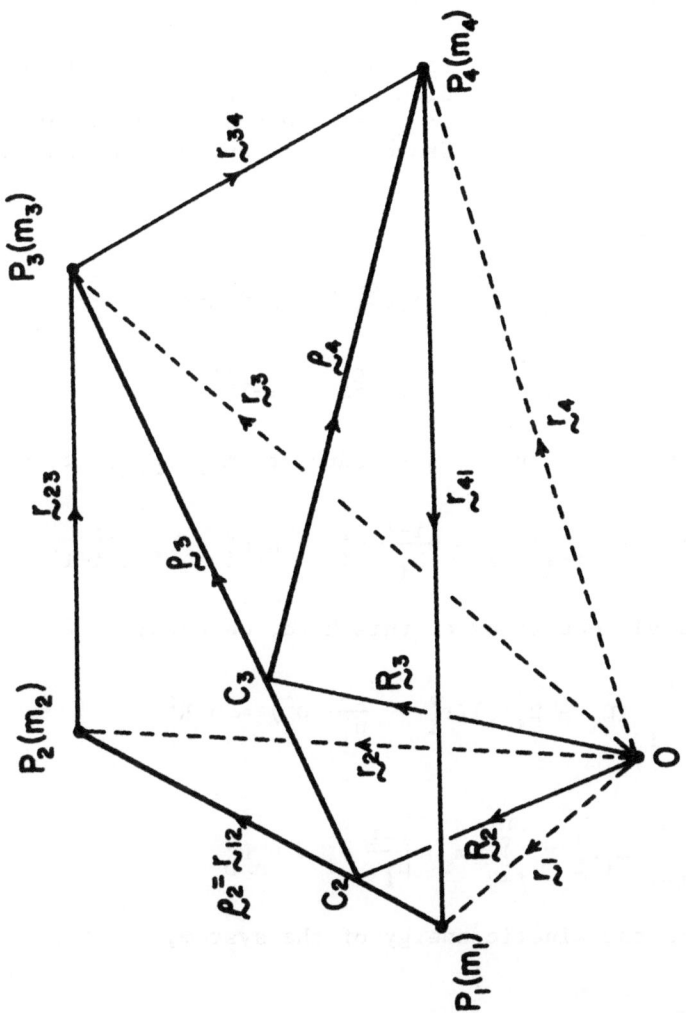

Figure 1

Then

$$\rho_i = r_i - R_{i-1},$$

and

$$(\mu_i - \mu_{i-1}) \rho_i = \mu_i (R_i - R_{i-1}). \tag{3}$$

In particular, ρ_2 is the radius vector of P_2 with respect to C_1, i.e., P_1; ρ_3 is the radius vector of P_3 with respect to C_2, centre of mass of P_1 and P_2; ρ_4 is the radius vector of P_4 with respect to C_3, centre of mass of P_1, P_2 and P_3; and so on. Note: ρ_1 does not exist.

Then by (2)

$$(\mu_i - \mu_{i-1})^2 r_i^2 = (\mu_i R_i^2 - \mu_{i-1} R_{i-1}^2)^2 \tag{4}$$

and

$$(\mu_i - \mu_{i-1})^2 \rho_i^2 = \mu_i^2 (R_i - R_{i-1})^2. \tag{5}$$

On eliminating the cross-product term $R_i \cdot R_{i-1}$, we have

$$(\mu_i - \mu_{i-1})(r_i^2 - \frac{\mu_{i-1}}{\mu_i} \rho_i^2) = \mu_i R_i^2 - \mu_{i-1} R_{i-1}^2. \tag{6}$$

Adding all equations of this kind, we obtain

$$\sum_{i=1}^{n} (\mu_i - \mu_{i-1})(r_i^2 - \frac{\mu_{i-1}}{\mu_i} \rho_i^2) = \mu_n R_n^2$$

or

$$\sum_{i=1}^{n} m_i r_i^2 = \sum_{i=2}^{n} m_i \frac{\mu_{i-1}}{\mu_i} \rho_i^2 + \mu_n R_n^2. \tag{7}$$

If T is the kinetic energy of the system,

$$T = \frac{1}{2} \sum_{i=1}^{n} m_i \dot{r}_i \cdot \dot{r}_i$$

so that

$$T = \frac{1}{2} \sum_{i=2}^{n} m_i \frac{\mu_{i-1}}{\mu_i} \dot{\rho}_i \cdot \dot{\rho}_i + \mu_n \dot{R}_n \cdot \dot{R}_n, \tag{8}$$

since the relation among the radius vectors are linear and therefore hold for derivatives as well.

Now $\underline{R}_n$ is the radius vector of the centre of mass of the system of bodies. It does not appear in the force function U given by

$$U = \frac{1}{2} G \sum_{i=1}^{n} \sum_{j=1}^{n} \frac{m_i m_j}{r_{ij}}, \quad j \neq i. \tag{9}$$

Hence $\underline{R}_n$ is ignorable. We may take the origin to be C_n and $\underline{R}_n = \dot{\underline{R}}_n = 0.$

We may now compute the remaining integrals of the system. The Hamiltonian H for the system is given by

$$H = T - U.$$

For this system H is constant and the integral of energy is

$$N = T - U = E, \tag{10}$$

or

$$\frac{1}{2} \sum_{i=2}^{n} m_i \frac{\mu_{i-1}}{\mu_i} \dot{\underline{\rho}}_i \cdot \dot{\underline{\rho}}_i - \frac{1}{2} G \sum_{i=1}^{n} \sum_{j=1}^{n} \frac{m_i m_j}{r_{ij}} = E. \tag{11}$$

The integral of angular momentum in Jacobian coordinates may be derived as follows:

Using (2), we have

$$(\mu_i - \mu_{i-1})^2 \underline{r}_i \times \dot{\underline{r}}_i = (\mu_i \underline{R}_i - \mu_{i-1} \underline{R}_{i-1}) \times (\mu_i \dot{\underline{R}}_i - \mu_{i-1} \dot{\underline{R}}_{i-1} \times \dot{\underline{R}}_{i-1} \tag{12}$$

From (3) we can write

$$(\mu_i - \mu_{i-1})^2 \underline{\rho}_i \times \dot{\underline{\rho}}_i = \mu_i^2 (\underline{R}_i - \underline{R}_{i-1}) \times (\dot{\underline{R}}_i - \dot{\underline{R}}_i). \tag{13}$$

Hence

$$(\mu_i - \mu_{i-1})(\underline{r}_i \times \dot{\underline{r}}_i - \frac{\mu_{i-1}}{\mu_i} \underline{\rho}_i \times \dot{\underline{\rho}}_i) = \mu_i \underline{R}_i \times \dot{\underline{R}}_i - \mu_{i-1} \underline{R}_{i-1} \times \dot{\underline{R}}_{i-1}. \tag{14}$$

Summing all equations of type (14), we obtain

$$\sum_{i=1}^{n} m_i (\underline{r}_i \times \dot{\underline{r}}_i - \frac{\mu_{i-1}}{\mu_i} \underline{\rho}_i \times \dot{\underline{\rho}}_i) = \mu_n \underline{R}_n \times \dot{\underline{R}}_n = 0.$$

Hence we have

$$\sum_{i=2}^{n} m_i \frac{\mu_{i-1}}{\mu_i} \underline{\rho}_i \times \underline{\dot{\rho}}_i = \underline{h},$$ (15)

where $\underline{h}$ is the constant angular momentum vector of the system with respect to the centre of mass. Relation (15) gives three components:

$$\left. \begin{aligned} \sum_{i=2}^{n} m_i \frac{\mu_{i-1}}{\mu_i} (x_i \dot{y}_i - y_i \dot{x}_i) &= h_z, \\[2ex] \sum_{i=2}^{n} m_i \frac{\mu_{i-1}}{\mu_i} (y_i \dot{z}_i - z_i \dot{y}_i) &= h_x, \\[2ex] \sum_{i=2}^{n} m_i \frac{\mu_{i-1}}{\mu_i} (z_i \dot{x}_i - x_i \dot{z}_i) &= h_y, \end{aligned} \right\}$$ (16)

where $\underline{\rho}_i = \underline{i} x_i + \underline{j} y_i + \underline{k} z_i$, $\underline{i}$, $\underline{j}$, $\underline{k}$ being unit vectors parallel to fixed rectangular axes $C_n x$, $C_n y$, $C_n z$ through the centre of mass, h_x, h_y, h_z are the components of $\underline{h}$, such that

$$\underline{h} = \underline{i} h_x + \underline{j} h_y + \underline{k} h_z.$$

Note that the axes $C_n x$, $C_n y$, $C_n z$ are parallel respectively to the axes $O\xi$, $O\eta$, $O\zeta$.

2.   THE EQUATIONS OF MOTION

We now set up the equations of motion of the bodies in the Jacobian system.

The Hamiltonian of the problem is given by

$$H = T - U = constant = E,$$ (20)

where   neglecting $\underline{R}_n$, as we may,

$$T = \frac{1}{2} \sum_{i=2}^{n} \frac{\mu_{i-1}}{\mu_i} \underline{\dot{\rho}}_i \cdot \underline{\dot{\rho}}_i$$ (21)

and

$$U = \frac{1}{2} G \sum_{i=1}^{n} \sum_{j=1}^{n} \frac{m_i m_j}{r_{ij}} , j \neq 1.$$ (22)

Now

$$(\mu_i - \mu_{i-1}) \, \underline{r}_i = \mu_i \underline{R}_i - \mu_{i-1} \underline{R}_{i-1}$$

$$= \mu_i (\underline{r}_{i+1} - \underline{\rho}_{i+1}) - \mu_{i-1} (\underline{r}_i - \underline{\rho}_i)$$

so that

$$\underline{r}_{i+1} - \underline{r}_i = \underline{\rho}_{i+1} - \frac{\mu_{i-1}}{\mu_i} \underline{\rho}_i . \qquad (23)$$

Summing equations of the form (23) we get

$$\left. \begin{array}{l} \underline{r}_{1,i+1} = \underline{r}_{i+1} - \underline{r}_1 = \underline{\rho}_{i+1} + \dfrac{m_i}{\mu_i} \underline{\rho}_i + \dfrac{m_{i-1}}{\mu_{i-1}} \underline{\rho}_{i-1} + \ldots + \dfrac{m_2}{\mu_2} \underline{\rho}_2 . \\[18pt] \text{In particular} \\[10pt] \underline{r}_{12} = \underline{r}_2 - \underline{r}_1 = \underline{\rho}_2 . \end{array} \right\} \qquad (24)$$

Hence r_{ij} in (22) may be expressed in terms of the vectors $\underline{\rho}_i$ or their components x_i, y_i, z_i.

Now define a momentum vector $\underline{p}_k$, where

$$\underline{p}_k = (\underline{i} \, \frac{\partial}{\partial \dot{x}_k} + \underline{j} \, \frac{\partial}{\partial \dot{y}_k} + \underline{k} \, \frac{\partial}{\partial \dot{z}_k}) T, \qquad (25)$$

Remembering that

$$\underline{\rho}_k = \underline{i} \, x_k + \underline{j} \, y_k + \underline{k} \, z_k .$$

Then

$$\underline{p}_k = m_k \, \frac{\mu_{k-1}}{\mu_k} \, (\underline{i} \, \dot{x}_k + \underline{j} \, \dot{y}_k + \underline{k} \, \dot{z}_k) = m_k \, \frac{\mu_{k-1}}{\mu_k} \, \dot{\underline{\rho}}_k$$

Hence, if x_i', y_i', z_i' are the components of vector $\underline{p}_i$, such that

$$\underline{p}_i = \underline{i} \, x_i' + \underline{j} \, y_i' + \underline{k} \, z_i' , \qquad (27)$$

we may write

$$x_i' = m_i \, \frac{\mu_{i-1}}{\mu_i} \, \dot{x}_i, \quad y_i' = m_i \, \frac{\mu_{i-1}}{\mu_i} \, \dot{y}_i, \quad z_i' = m_i \, \frac{\mu_{i-1}}{\mu_i} \, \dot{z}_i \quad (28)$$

The Hamiltonian canonical equations of the dynamical system now may be written in the usual way, viz,

$$\dot{\rho}_i = \frac{\partial}{\partial p_i} \quad , \quad \dot{p}_i = -\frac{\partial}{\partial \rho_i} ,$$

where

$$\frac{\partial}{\partial p_i} = \underline{i}\,\frac{\partial}{\partial x_i'} + \underline{j}\,\frac{\partial}{\partial y_i'} + \underline{k}\,\frac{\partial}{\partial z_i'} ,$$

and

$$\frac{\partial}{\partial \rho_i} = \underline{i}\,\frac{\partial}{\partial x_i} + \underline{j}\,\frac{\partial}{\partial y_i} + \underline{k}\,\frac{\partial}{\partial z_i} .$$

$$(29)$$

3.   GENERAL PERTURBATION DEVELOPMENT IN THE PLANETARY CASE

Suppose we are dealing with the solar system and we identify P_1 with the sun. Then $m_i \ll m_1$, $i > 1$. By (23) and (24) we

$$\underline{r}_{i+1} - \underline{r}_i = \underline{\rho}_{i+1} - \frac{\mu_{i-1}}{\mu_i}\,\underline{\rho}_i ,$$

$$\underline{r}_{1,i+1} = \underline{r}_{i+1} - \underline{r}_1 = \underline{\rho}_{i+1} + \frac{m_i}{\mu_i}\,\underline{\rho}_i + \ldots + \frac{m_2}{\mu_2}\,\underline{\rho}_2 ;$$

$$\underline{r}_{12} = \underline{r}_2 - \underline{r}_1 = \underline{\rho}_2 ,$$

showing that in the planetary case the relative vectors $\underline{r}_{ij}$ differ from the $\underline{\rho}$ vectors only by quantities of the first order in the planetary masses. It is noted also that for the body P_2, chosen as any one of the planets, there is no difference between $\underline{r}_{12}$ and $\underline{\rho}_2$.

The zero order solution is obtained by restricting the force function U to the lowest order in the small masses. We have

$$U_o = G\,m_1 \sum_{i=2}^{n} \frac{m_i}{\rho_i}$$

$$(30)$$

where $\rho_i^2 = x_i^2 + y_i^2 + z_i^2$, since $\underline{\rho}_i$ differs from $\underline{r}_{ij}$ by a quantity which involves the planetary masses. Then the Hamiltonian for this approximation is

$$H_o = T - U_o.$$

We have canonical equations given by (29), viz,

$$\dot{\rho}_i = \frac{\partial H_o}{\partial p_i} \; ; \quad \dot{p}_i = - \frac{\partial H_o}{\partial \rho_i} .$$

These give, on substitution,

$$\frac{\mu_{i-1}}{\mu_i} \ddot{\rho}_i = - Gm_1 \frac{\rho_i}{\rho_i^3} , \quad i=2,3,\ldots,n.$$

For example, putting i=2, we obtain

$$\frac{\mu_1}{\mu_2} \ddot{\rho}_2 = - Gm_1 \frac{\rho_2}{\rho_2^3} ,$$

or, remembering that $\mu_1 = m_1$, $\mu_2 = m_1 + m_2$, we have

$$\ddot{\rho}_2 + G(m_1 + m_2) \frac{\rho_2}{\rho_2^3} = 0 ,$$

the solution for the unperturbed elliptic motion of planet P_2 about the Sun P_1.

It should be noted, however, that for other values of i, namely 3 and upwards, although the equations of motion are of the same form and give similar solutions, the elements will not agree with the usual ones since each ρ_i, i > 2 is a vector drawn from the center of mass C_{i-1} to P_i.

The canonical constants obtained from the solution of the Hamiltonian canonical equations (29) using H_o, can now be treated in the usual way as new canonical variables satisfying Hamilton's canonical equations using $H_1 = H - H_o$ as the new Hamiltonian. As is well known, the process of obtaining higher order perturbations may in principle be arbitrarily extended.

This general perturbation development is a process hallowed in the annals of celestial mechanics leading to such discoveries that to the second order in the masses there is no secular inequality in the mean distances of the planetary orbits (Poisson's theorem which is an extension of Laplace's theorem for the first order). Hailed as a proof of the stability of the solar system it was left to Poincaré to point out that in fact it says nothing about the long-term stability of the solar system since there is proof that the series obtained are not uniformly convergent.

4.   THE GENERAL FOUR-BODY PROBLEM

It will be of use to set up explicitly the equations of motion of the four-body problem in Jacobian coordinates.

We have $H = T - U$, where in this case

$$T = \frac{1}{2} \sum_{i=2}^{4} m_i \frac{\mu_{i-1}}{\mu_i} \dot{\rho}_i \cdot \dot{\rho}_i ,$$

and

$$U = \frac{1}{2} G \sum_{i=1}^{4} \sum_{j=1}^{4} \frac{m_i m_j}{r_{ij}} , \quad j \neq i.$$

The equations of motion, obtained from (29) are

$$m_i \frac{\mu_{i-1}}{\mu_i} \ddot{\rho}_i = \frac{\partial U}{\partial \rho_i} . \tag{31}$$

By (24) we have

$$r_{12} = \rho_2 ; \quad r_{13} = \rho_3 + \frac{m_2}{\mu_2} \rho_2; \quad r_{14} = \rho_4 + \frac{m_3}{\mu_3} \rho_3 + \frac{m_2}{\mu_2} \rho_2;$$

$$r_{23} = \rho_3 - \frac{m_1}{\mu_2} \rho_2 ; \quad r_{24} = \rho_4 + \frac{m_3}{\mu_3} \rho_3 - \frac{m_1}{\mu_2} \rho_2;$$

$$r_{34} = \rho_4 - \frac{\mu_2}{\mu_2} \rho_3.$$

Hence by (31) we obtain

$$\ddot{\rho}_2 + G\mu_2 \frac{\rho_2}{2} = G \frac{\mu_2}{m_1 m_2} \left\{ m_3 \frac{\partial}{\partial \rho_2} \left[\frac{m_1}{r_{13}} + \frac{m_2}{r_{23}} \right] \right.$$

$$\left. + m_4 \frac{\partial}{\partial \rho_2} \left[\frac{m_1}{r_{14}} + \frac{m_2}{r_{24}} \right] \right\} . \tag{32}$$

Similarly, for i=3, we obtain

$$\ddot{\rho}_3 = \frac{G\mu_2}{m_3 \mu_2} \left\{ m_3 \frac{\partial}{\partial \rho_3} \left[\frac{m_1}{r_{13}} + \frac{m_2}{r_{23}} \right] + m_4 \frac{\partial}{\partial \rho_3} \left[\frac{m_1}{r_{14}} + \frac{m_2}{r_{24}} + \frac{m_3}{r_{34}} \right] \right\}.$$

$$\tag{33}$$

For i=4, we have

$$\ddot{\underline{\rho}}_3 = \frac{G\mu_4}{\mu_3} \frac{\partial}{\partial\underline{\rho}_4} \left(\frac{m_1}{r_{14}} + \frac{m_2}{r_{24}} + \frac{m_3}{r_{34}} \right) . \tag{34}$$

Introducing potential functions B and F, we may rewrite these equations in the form

$$\ddot{\underline{\rho}}_2 + G\mu_2 \frac{\underline{\rho}_2}{\rho_2^3} = G \frac{\mu_2}{m_1 m_2} \left(m_3 \frac{\partial B}{\partial\underline{\rho}_2} + m_4 \frac{\partial F}{\partial\underline{\rho}_2} \right) , \tag{35}$$

$$\ddot{\underline{\rho}}_3 \qquad = G \frac{\mu_3}{m_3 \mu_2} \left(m_3 \frac{\partial B}{\partial\underline{\rho}_3} + m_4 \frac{\partial F}{\partial\underline{\rho}_3} \right) \tag{36}$$

$$\ddot{\underline{\rho}}_4 \qquad = G \frac{\mu_4}{\mu_3} \frac{\partial F}{\partial\underline{\rho}_4} , \tag{37}$$

where

$$B = \frac{m_1}{r_{13}} + \frac{m_2}{r_{23}} , \qquad F = \frac{m_1}{r_{14}} + \frac{m_2}{r_{24}} + \frac{m_3}{r_{34}} ,$$

since, in (35),

$$\frac{\partial}{\partial\underline{\rho}_2} \left(\frac{m_3}{r_{34}} \right) = 0 .$$

In order to illustrate the philosophy to be followed, let us first consider the general three-body problem.

5.   THE GENERAL THREE-BODY PROBLEM

There is no equation in $\underline{\rho}_4$ and the equations (35) and (36) reduce to

$$\ddot{\underline{\rho}}_2 = G\mu_2 \frac{\partial}{\partial\underline{\rho}_2} \left[\frac{1}{\rho_2} + \frac{m_3}{m_1 m_2} B \right] , \tag{38}$$

$$\ddot{\underline{\rho}}_3 = G \frac{\mu_3}{\mu_2} \frac{\partial B}{\partial\underline{\rho}_3} . \tag{39}$$

These equations are the well known equations of motion for the lunar problem or the stellar three-body problem.

Define $\alpha = \rho_2/\rho_3$, $c = \cos\theta$ where θ is the angle between $\underline{\rho}_2$ and $\underline{\rho}_3$ (see Figure 1). Let $q_1 = m_1/\mu_2 \, \rho_2$, $q_2 = m_2/\mu_2 \, \rho_2$, so that we have

$$\frac{\rho_3}{r_{23}} = [1 + (q_1/\rho_3)^2 - 2(q_1 c/\rho_3)]^{-1/2}.$$

If $\alpha <$ we may expand this in Legendre polynomials, obtaining

$$\frac{\rho_3}{r_{23}} = 1 + \sum_{j=1}^{\infty} \alpha_1^j \, P_j(c). \tag{40}$$

where $\alpha_1 = q_1/\rho_3$.

Similarly, putting $\alpha_2 = q_2/\rho_3$, we obtain

$$\frac{\rho_3}{r_{13}} = 1 + \sum_{j=1}^{\infty} (-\alpha_2)^j \, P_j(c). \tag{41}$$

Then $B = (1/\rho_3) \, [m_1 (\rho_3/r_{13}) + m_2 \, (\rho_3/r_{23})]$ is given by the well known expression

$$B = \frac{\mu_2}{\rho_3} + \frac{m_1 m_2}{\mu_2} \left[\frac{\rho_2^3}{\rho_3^3} P_2 + (\frac{m_1 - m_2}{m_1 + m_2}) \frac{\rho_2^3}{\rho_3^4} P^3 \right.$$
$$\left. + (\frac{m_1^2 - m_1 m_2 + m_2^2}{(m_1 + m_2)^2}) \frac{\rho_2^4}{\rho_3^5} P_4 + \dots \right]. \tag{42}$$

Substituting B from (42) into (38), we obtain

$$\ddot{\underline{\rho}}_2 = G\mu_2 \frac{\partial}{\partial \underline{\rho}_2} \left\{ \frac{1}{\rho_2} + \frac{m_3}{\mu_2} \left[\frac{\rho_2^2}{\rho_3^3} P_2 + (\frac{m_1 - m_2}{m_1 + m_2}) \frac{\rho_2^3}{\rho_3^4} P_3 + \dots \right] \right\} \tag{43}$$

Note that the term ρ_3^{-1} in B may be neglected in (43). Substituting for B in (39) we have

$$\ddot{\underline{\rho}}_3 = G\mu_3 \frac{\partial}{\partial\underline{\rho}_3} \left\{ \frac{1}{\rho_3} + \frac{m_1 m_2}{\mu_2^2} \left[\frac{\rho_2^2}{\rho_3^3} P_2 + (\frac{m_1 - m_2}{m_1 + m_2}) \frac{\rho_2^3}{\rho_3^2} P_3 + \ldots \right] \right\} \quad (44)$$

Now it is the first term inside the curly brackets of (43) and (44) that would give undisturbed elliptic motion if all other terms on the right hand sides were neglected, firstly of mass m_2 about mass m_1 and secondly of mass m_3 about the centre of mass m_1 and m_2. The second and subsequent terms provide the perturbation.

Define

$$\left. \begin{array}{l} \varepsilon_3 = \frac{m_3}{m_1 + m_2} \left(\frac{\rho_2}{\rho_3} \right)^3 = \frac{m_3}{m_1 + m_2} \alpha^3 \\[2mm] \varepsilon_{12} = \frac{m_1 m_2}{(m_1 + m_2)^2} \left(\frac{\rho_2}{\rho_3} \right)^2 = \frac{m_1 m_2}{(m_1 + m_2)^2} \alpha^2 \end{array} \right\} \quad (45)$$

where $\alpha = \rho_2/\rho_3$. Both ε_3 and ε_{12} are non-dimensional variables and may be taken as a measure of the disturbance of m_3 on the $m_1 m_2$ binary and of the disturbance of m_1 and m_2 on the orbit of m_3 about the centre of mass of m_1 and m_2.

Now $\alpha \leq 1$, hence $0 \leq \varepsilon_{12} \leq 0.25$, while in theory $0 \leq \varepsilon_3 \leq \varepsilon$ where ε may be arbitrarily large. In practice, however, it is obvious that ε_3 must lie in the range $0 \leq \varepsilon_3 \leq 1$ if any kind of stability is to be possible.

Examples

(1) Triple stellar system with $m_1 = m_2 = m_3 = m$.

Then

$$\varepsilon_3 = 0.5\alpha^3 \quad ; \quad \varepsilon_{12} = 0.25\alpha^2 .$$

(2) Earth-Moon-Sun system with $m_1 = 1$, $m_2 = 0.0123$, $m_3 = 330,000$.

Then

$$\varepsilon_3 \simeq 3.3 \times 10^5 \alpha^3 \quad ; \quad \varepsilon_{12} \simeq 1.23 \times 10^{-2} \alpha^2 .$$

(3) Sun-Jupiter-Saturn with $m_1 = 1$, $m_2 = 10^{-3}$, $m_3 = 3 \times 10^{-4}$.

Then

$$\varepsilon_3 \simeq 3 \times 10^{-4} \alpha^3 \quad ; \quad \varepsilon_{12} \simeq 10^{-3} \alpha^2 .$$

For any real three–body system in nature, if α is known as well as the masses, then values of the two ε's may be computed. In fact if we define

$$\mu = \frac{m_2}{m_1 + m_2} \; , \quad \mu_3 = \frac{m_3}{m_1 + m_2} \; , \quad \text{it is seen that}$$

any three–body system may be reduced to a three parameter system, either a set (μ, μ_3, α) or a set $(\varepsilon_{12}, \varepsilon_3, \alpha)$. The eccentricity and inclination parameters, also the question of direct and retro-grade motion and the effect of commensurability in mean motion are neglected meantime.

6.   DEVELOPMENT OF THE FOUR–BODY EQUATIONS OF MOTION

We require the expansions of

$$B = \frac{m_1}{r_{13}} + \frac{m_2}{r_{23}} \; , \quad F = \frac{m_1}{r_{14}} + \frac{m_2}{r_{24}} + \frac{m_3}{r_{34}} \; .$$

The four–body system is not necessarily coplanar so that we define

$$c_{ij} = \cos\theta_{ij} = \frac{\rho_i \cdot \rho_j}{\rho_i \, \rho_j} \; .$$

Then B is given by

$$B = \frac{\mu_2}{P_3} + \frac{m_1 m_2}{\mu_2} \left[\frac{\rho_2^2}{\rho_3^3} P_2(c_{23}) + \left(\frac{m_1 - m_2}{m_1 + m_2}\right) \frac{\rho_2^3}{\rho_3^4} P_3(c_{23}) + \cdots \right. \; .$$

Now

$$r_{14} = \left| \rho_4 + \frac{m_3}{\mu_3} \rho_3 + \frac{m_2}{\mu_2} \rho_2 \right| \; , \quad r_{24} = \left| \rho_4 + \frac{m_3}{\mu_3} \rho_3 = \frac{m_1}{\mu_2} \rho_2 \right| \; ,$$

$$r_{34} = \left| \rho_4 - \frac{\mu_2}{\mu_3} \rho_3 \right| \; .$$

After some reduction, we find that

$$F = \frac{1}{\rho_4} \left[\mu_3 + \frac{m_1 m_2}{\mu_2} \left(\frac{\rho_2}{\rho_4}\right)^2 P_2(c_{24}) + \frac{m_1 \mu_2}{\mu_3} \left(\frac{\rho_3}{\rho_4}\right)^2 P_2(c_{34}) + \cdots \right] \; .$$

Taking equations (35), (36) and (37) and substituting the above expressions for B and F, we obtain

$$\ddot{\underline{\rho}}_2 = G\mu_2 \frac{\partial}{\partial\underline{\rho}_2} \left\{ \left[\frac{1}{\rho_2} + \frac{m_3}{m_1 m_2 \rho_3} \frac{2}{} + \frac{m_3}{\mu_3} \frac{\rho_2^2}{\rho_3^3}) P_2(c_{23}) + \dots \right] + \right.$$
$$\left. + \frac{m_4}{m_1 m_2} \left[\frac{\mu_3}{\rho_4} + \frac{m_1 m_2}{\rho_2} \frac{\rho_2^2}{\rho_3^3}) P_2(c_{24}) + \frac{m_3 \mu_2}{\mu_3} (\frac{\rho_3^2}{\rho_4^3}) P_2(c_{34}) + \dots \right] \right\}$$
(46)

$$\ddot{\underline{\rho}}_3 = G \frac{\mu_3}{m_3 \mu_2} \frac{\partial}{\partial\rho_3} \left\{ \left[\frac{m_3 \mu_2}{\rho_3} + \frac{m_1 m_2 m_3}{\mu_2} (\frac{\rho_2^2}{\rho_3^3}) P_2(c_{23}) + \dots \right] + \right.$$
$$\left. + m_4 \left[\frac{\mu_3}{\rho_4} + \frac{m_1 m_2}{\mu_2} (\frac{\rho_2^2}{\rho_4^3}) P_2(c_{24}) + \frac{m_3 \mu_2}{\mu_3} (\frac{\rho_3^2}{\rho_4^3}) P_2(c_{34}) + \dots \right] \right\}$$
(47)

$$\ddot{\underline{\rho}}_4 = \frac{G\mu_4}{\mu_3} \frac{\partial}{\partial\rho_4} \left\{ \frac{\mu_3}{\rho_4} + \frac{m_1 m_2}{\mu_2} (\frac{\rho_2^2}{\rho_3^3}) P_2(c_{24}) + \frac{m_3 \mu_2}{\mu_3} (\frac{\rho_3^2}{\rho_4^3}) P_2(c_{34}) + \dots \right\}$$
(48)

Examining (46) through (48) it is seen that the first term on the right hand sides gives the undisturbed elliptic motion. Now consider the ratios of the second non-zero terms with the first in each equation within the curly brackets. It is clear that we can define nondimensional quantities thus:

From equation (46):

$$\varepsilon_3 = \frac{m_3}{\mu_2} (\frac{\rho_2}{\rho_3})^3 \qquad ; \qquad \varepsilon_4 = \frac{m_4}{\mu_2} (\frac{\rho_2}{\rho_4})^3 .$$

From equation (47):

$$\varepsilon_3' = \frac{m_1 m_2}{\mu_2^2} (\frac{\rho_2}{\rho_3})^2 \qquad ; \qquad \varepsilon_4' = \frac{m_4}{\mu_2} (\frac{\rho_3}{\rho_4})^3 .$$

From equation (48):

$$\varepsilon_3'' = \frac{m_1 m_2}{\mu_2 \mu_3} (\frac{\rho_2}{\rho_4})^2 \qquad ; \qquad \varepsilon_4'' = \frac{m_3 \mu_2}{3} (\frac{\rho_3}{\rho_4})^2 .$$

If we let $\alpha = \rho_2/\rho_3$, $\beta = \rho_3/\rho_4$, $\gamma = \rho_2/\rho_4 = \alpha\beta$, we are ready to consider some examples.

1   Sun–Jupiter–Saturn–Uranus. Here

$$m_1 = 1, \ m_2 = 10^{-3}, \ m_3 = 3.10^{-4}, \ m_4 = 5.10^{-5} ;$$

$$\rho_2 = 5.2, \ \rho_3 = 9.5, \ \rho_4 = 19.2.$$

Then $\alpha = 0.55, \ \beta = 0.49, \ \gamma = 0.27$, and we find that

$$\varepsilon_3 \simeq 5.10^{-5} ; \quad \varepsilon_4 \simeq 10^{-6}$$

$$\varepsilon_3' \simeq 3.10^{-4} ; \quad \varepsilon_4' \simeq 6.10^{-6}$$

$$\varepsilon_3'' \simeq 7.10^{-5} ; \quad \varepsilon_4'' \simeq 7.10^{-5} .$$

A measure of the combined perturbations of Saturn and Uranus on Jupiter is given by $\varepsilon_3 + \varepsilon_4 \simeq 5.1 \times 10^{-5}$. Similarly, Jupiter and Uranus' perturbations on Saturn is given as $\varepsilon_3' + \varepsilon_4' \simeq 3.06 \times 10^{-4}$, while Jupiter and Saturn's perturbations on Uranus are of the order $\varepsilon_3'' + \varepsilon_4'' \simeq 1.4 \times 10^{-4}$.

2   Four-star system in a hierarchy where m_3 revolves about C_2 while m_4 revolves about C_3. If all masses are taken equal then,

$$\varepsilon_3 = 0.5\alpha^3 \quad , \quad \varepsilon_4 = 0.5\gamma^3$$

$$\varepsilon_3' = 0.25\alpha^2 \quad , \quad \varepsilon_4' = 0.33\beta^3$$

$$\varepsilon_3'' = 0.16\gamma^2 \quad , \quad \varepsilon_4'' = 0.22\beta^2 .$$

If $\alpha = \beta = 0.1$, so that $\gamma = 10^{-2}$, the values of the epsilons are suitably small enough to ensure no immediate disruption of the system. The value 0.1 is not untypical of such stellar systems.

It is obvious that the n-body problem, $n \geq 3$, expressed in the Jacobian coordinate system, can have epsilons defined for it such that any equation may be written in the form

$$\ddot{\rho}_i = G \frac{\partial}{\partial \rho_i} \left[\frac{\mu_i}{\rho_i} (1 + \sum_j \varepsilon_{ij} W_{ij}) \right] ,$$

or

$$\ddot{\rho}_i + G\mu_i \frac{\rho_i}{\rho_i} = G \frac{\partial}{\partial \rho_i} \left[\frac{\mu_i}{\rho_i} \sum_j \varepsilon_{ij} W_{ij} \right] , \tag{49}$$

where the W_{ij} are functions of the angles between the Jacobian vectors.

We now <u>conjecture</u> that if all the equations in an n-body problem of this kind (i.e., as possessing a hierarchical structure) have

$$\sum_j |\varepsilon_{ij}| < \ell ,$$

where ℓ is some value much less than unity, the solutions to the equations will be stable in the sense that the order of the bodies will not be changed for an astronomically long time, for example, a satellite will not escape or two planets reverse their positions.

Is there any evidence to support this conjecture?

In what follows we consider firstly the relevance of the work of Zare, Szebehely and McKenzie on the question of stability in the general three-body problem.

7. THE CRITICAL STABILITY SURFACE IN THE GENERAL COPLANAR THREE-BODY PROBLEM

Recently Zare (1976,1977), Szebehely (1977), Szebehely and Zare (1977), Szebehely and McKenzie (1977a,b) have extended the concept of surfaces of zero velocity and their relevance to stability in the restricted three-body problem to the general coplanar three-body problem. A condition has been derived sufficient for stability in the sense that if two of the masses m_1 and m_2 are in orbit about each other, the third being in orbit about the first two's centre of mass, there can be no break-up of the $m_1 m_2$ binary if the condition is satisfied. This condition is found to be consistent with various numerical studies of stability in three-body dynamical systems: see for example Nacozy (1976) and Szebehely and McKenzie (1977a).

Zare (1976,1977) showed that stability was controlled by values of a parameter $S = C^2 E$, where C is the angular momentum of the system and E its total energy. Szebehely and Zare (1977) have presented a concise summary of the necessary steps in the problem of computing a measure of the stability of any coplanar general three-body system.

The criterion for stability is most conveniently expressed in terms of the parameter s, defined by

$$s = S - S_{cr}/S_{cr} , \qquad (51)$$

where S_{cr} is the critical value of C^2E for the system. For $s > 0$ the system is stable in that no exchange between masses can occur, whereas for $s < 0$, exchange may or may not occur. Hence the critical boundary between $s > 0$ and $s < 0$ occurs for $S = S_{cr}$. The critical value of S, namely S_{cr}, is evaluated by first solving the appropriate and well-known quintic equation in μ and μ_3, where

$$\mu = \frac{m_2}{m_1 + m_2} \quad , \quad \mu_3 = \frac{m_3}{m_1 + m_2} \quad . \tag{52}$$

The quintic arises from the Lagrangian collinear solution of the three-body problem. There are three such solutions depending upon the ordering of the masses but one (m_1, m_2, m_3) can immediately be discarded. Of the other two, viz (m_3, m_1, m_2) and (m_3, m_2, m_1), the choice depends on whether or not m_3 is the smallest mass. If not, we use m_3, $\min(m_1, m_2)$, $\max(m_1, m_2)$ to give the largest critical value at the primary bifurcation. If m_3 is the smallest mass, the secondary bifurcation determines S_{cr}. The one positive root ρ of the quintic then enables two relations, namely

$$f(\rho) = \mu\mu_3 + \frac{(1-\mu)\mu_3}{1 + \rho} + \frac{\mu(1-\mu)}{\rho} \tag{53}$$

and

$$g(\rho) = \mu\mu_3 + \mu_3(1-\mu)(1+\rho)^2 + \mu(1-\mu)\rho^2 \tag{54}$$

to be evaluated, where ρ is the distance $\overline{m_1 m_2}$ when $\overline{m_3 m_1} = 1$. Zare has shown that S_{cr} may then be computed from the expression

$$S_{cr} = - \frac{G^2 f^2(\rho) \ g(\rho)}{2(1+\mu_3)} \quad . \tag{55}$$

The actual value of S for the system is given by

$$-\frac{S}{G^2} = \frac{1}{2} \mu^2 (1-\mu)^2 \mu_3 (1-e_1^2)\alpha \pm \frac{\mu(1-\mu)\mu_3^2}{(1+\mu_3)^{1/2}} (1-e_1^2)^{\frac{1}{2}} (1-e_2^2)^{\frac{1}{2}} \alpha^{\frac{1}{2}}$$

$$+ \frac{1}{2} [\mu^3(1-\mu)^3 \ (1-e_1^2) + \frac{\mu_3^3}{1+\mu_3} (1-e_2^2)]$$

$$\pm \frac{\mu^2(1-\mu)^2\mu_3}{(1+\mu_3)^{1/2}} (1-e_1^2)^{\frac{1}{2}} (1-e_2^2)^{\frac{1}{2}} \ \alpha^{-\frac{1}{2}}$$

$$+ \frac{1}{2} \frac{\mu(1-\mu) \ \mu_3^2}{1 + \mu_3} (1-e_2^2)\alpha^{-1}$$

where e_1 and e_2 are the eccentricities in the binary and the external body's orbits, respectively, and α is the ratio of the semimajor axis of inner and outer orbits. The plus and minus signs refer to direct and retrograde motion of the outer component.

We can therefore compute for any given three-body system (i.e., for given values of μ and μ_3) the value of α, namely α_c, that gives critical stability (s=0). We call the surface $\alpha \overset{\cong}{=} \alpha_c$ (μ,μ_3) a surface of critical stability. The actual value of α for a given three-body system will give in the (μ,μ_3,α) space a point that in general lies above or below the critical stability surface. If it is above, instability may occur; if below, the system is stable. Obviously two critical stability surfaces exist, for direct and retrograde motion.

Although these surfaces are of decided interest (various cuts across them have been given in Szebehely and Zare (1977)), we shall recast the α-surface in terms of the epsilon parameters derived in previous sections, in particular ε_{12} and ε_3 defined in equation (45) to describe work, as yet unpublished, by Roy, Emslie and Walker.

Neglecting for the moment e_1 and e_2 in (56) we carry out the process of deriving the α-surface, that is, the critical stability surface in the $(\varepsilon_{12},\varepsilon_3,\alpha)$ space. For each pair of values of μ and μ_3, the critical stability value of α was computed, the pair of values of ε_{12} and ε_3 corresponding to these values of α, μ and μ_3 being found from

$$\varepsilon_{12} = \mu(1-\mu)\alpha^2, \quad \varepsilon_3 = \mu_3 \alpha^3.$$

The following ranges of the mass parameters μ and μ_3 were used, viz:

$$0 \leq \mu \leq 0.5, \quad 10^{-7} \leq \mu_3 \leq 10^7,$$

these ranges being chosen to include the six types of three-body systems listed below.

1. Two planets in orbit about the Sun, disturbing each other,
2a. A planet in orbit about one member of a binary star, perturbed by the other, more distant, member.
2b. A planet in orbit about the centre of mass of a binary star.
3a. A satellite in orbit about a planet, subject to solar perturbations.
3b. Two satellites in orbit about a planet, disturbing each other.
4. A triple stellar system.

In the cases found in nature, values of α occur in the range $10^{-4} < \alpha < 0.85$. The critical stability surface in μ, μ_3, α_c is shown for direct motion in Figure 2. The surface is so simple that it can be clearly plotted as contours in the μ, μ_3 plane (Figure 3). Now consider the distribution of all triple systems found in nature, as well as a number of numerical experiments in triple systems from the six cases listed above. Table 1 shows the available data for the solar system in terms of our knowledge of the number of orbits any body is known to have performed since its discovery. For many the number is greater than 10^3, giving an indication of their stability. Using data for semimajor axes and neglecting eccentricities and inclinations which are in general small, we process all triple systems found in nature, for example Sun-Moon-Earth, Sun-Jupiter-Saturn, Sun-asteroid-Jupiter, Sun-Venus-Earth, Jupiter-Io-Europa, Jupiter-Ganymede-Sun, and so on, including triple star systems. For each triple system a point is obtained in the μ, μ_3, α space. The projections of these points are given in Figure 3 where points denote examples of case 1, open circles denote asteroids disturbed by Jupiter, filled circles refer to numerical experiments of case 1, open squares are examples of case 2b, filled squares refer to case 2a, crosses refer to case 3a, pluses to case 3b, and asterisks to case 4.

It is found that even though i≠0 and even though the triple system itself is part of a larger system, very good agreement is obtained in that the vast majority of subsets of triples satisfy the critical stability criterion, that is, their α values lie well below the α_c surface. Even if the orbits' eccentricities were taken into account, they are mosly so small that the α_c surface should not be appreciably lowered. Those systems whose α's are greater than the corresponding α_c's are mainly those numerical experiments taken from Harrington (1977) and Horedt et al (1977), are known to be unstable.

The most interesting features emerge, however, if the data are transformed to the $\varepsilon_{12}, \varepsilon_3, \alpha$ space. If we construct the critical stability α_c surface (again for $e_1 = e_2 = 0$) the simple surface given in Figure 4 is obtained. There are also shown some limitations on possible values of α_c because of the definition of $\varepsilon_{12} = \mu(1-\mu)\alpha_c^2$ and the fact that μ must be real. Thus

$$\mu = \frac{1}{2} - \sqrt{\left(\frac{1}{4} - \frac{\varepsilon_{12}}{\alpha_c^2}\right)}, \quad 0 \le \mu \le \frac{1}{2}$$

so that μ is real only for $\frac{1}{4} - \frac{\varepsilon_{12}}{\alpha_c^2} \ge 0$.

The secondary surface such that α_c cannot fall below $2\varepsilon_{12}$ is also depicted in Figure 4, creating a "birdcage" within which three-body systems are stable (or trapped!) and outside of which

TABLE 1. Number of orbits performed by solar system objects since their discovery.

OBJECT	YEAR OF DISCOVERY	PERIOD years	NUMBER OF ORBITS OBSERVED SINCE DISCOVERY OR COMPUTED *	
Mercury	Before 2000 BC	0.241	1.7×10^4	
Venus	"	0.615	6.5×10^3	
Earth	"	1.000	4.0×10^3	
Mars	"	1.881	2.1×10^3	
Asteroids	1801	4.6	3.9×10^1	
Jupiter	Before 2000 BC	11.86	3.4×10^2	4.2×10^4*
Saturn	Before 2000 BC	29.46	1.3×10^2	1.7×10^4*
Uranus	1781	84.01	2.4	6.0×10^3*
Neptune	1846	164.78	1.0	3.0×10^3*
Pluto	1930	248.4	0.2	2.0×10^3*
		days		
Moon	Before 2000 BC	27.32	5.3×10^4	
Phobos	1877	0.319	1.2×10^5	
Deimos	1877	1.262	2.9×10^4	
Jupiter V	1892	0.498	6.3×10^4	
I	1610	1.769	7.6×10^4	
II	1610	3.551	3.8×10^4	
III	1610	7.155	1.9×10^4	
IV	1610	16.69	8.1×10^3	
VI	1904	251.	1.1×10^2	
VII	1905	260.	1.0×10^2	
X	1938	260.	5.6×10^1	
XIII	1975	252.	4.3	
VIII	1908	739.	3.5×10^1	
IX	1914	745.	3.1×10^1	
XI	1938	696.	2.1×10^1	
XII	1951	617.	1.0×10^1	
Janus	1966	0.749	5.9×10^3	
Mimas	1789	0.942	7.3×10^4	
Enceladus	1789	1.370	5.0×10^4	
Tethys	1684	1.888	5.7×10^4	
Dione	1684	2.737	3.9×10^4	
Rhea	1672	4.518	2.5×10^4	
Titan	1655	15.95	7.4×10^3	
Hyperion	1848	21.28	2.2×10^3	
Iapetus	1671	79.33	1.4×10^3	

* Cohen, Hubbard and Oesterwinter's numerical integration over 50,000 years (asterisk from preceeding page).

OBJECT	YEAR OF DISCOVERY	PERIOD days	NUMBER OF ORBITS OBSERVED SINCE DISCOVERY OR COMPUTED
Phoebe	1898	550.5	5.3×10^1
Miranda	1948	1.414	1.1×10^4
Ariel	1851	2.520	1.8×10^4
Umbriel	1851	4.144	1.1×10^4
Titania	1787	8.706	8.0×10^3
Oberon	1787	13.46	5.2×10^3
Triton	1846	5.877	8.2×10^3
Nereid	1949	359.	3.0×10^1

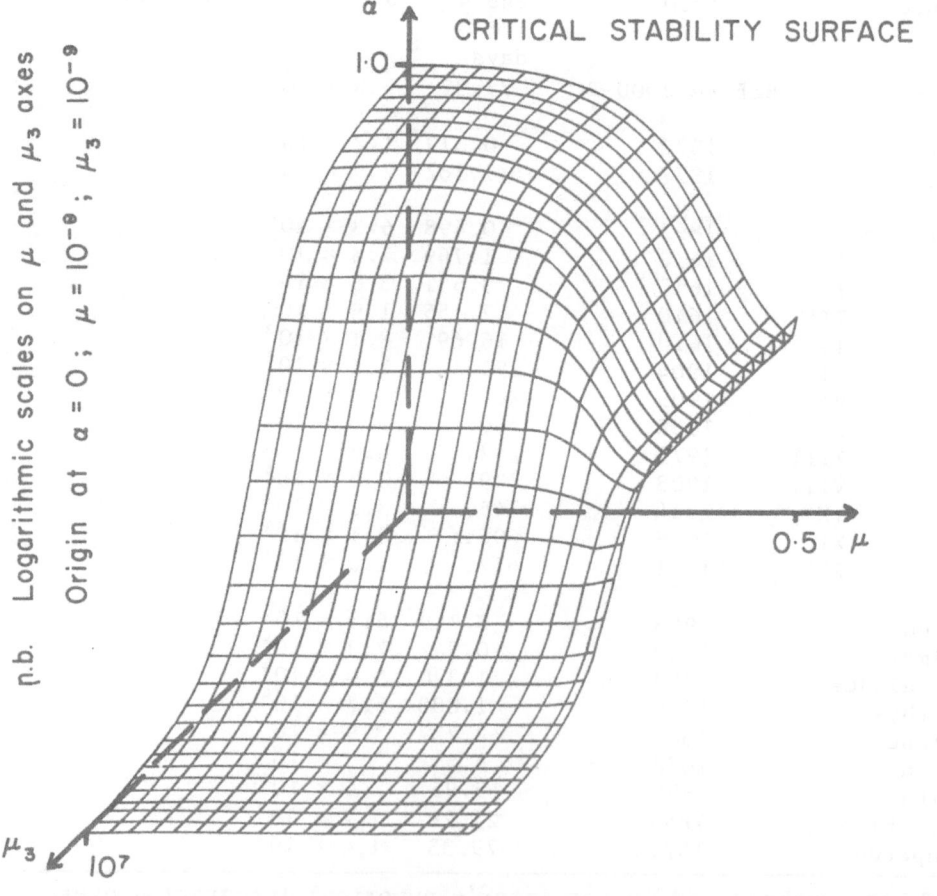

Figure 2.   Critical stability surface in the μ,μ_3,α space accord-
    ing to Zare's critical stability criterion in the general
    coplanar three-body problem.

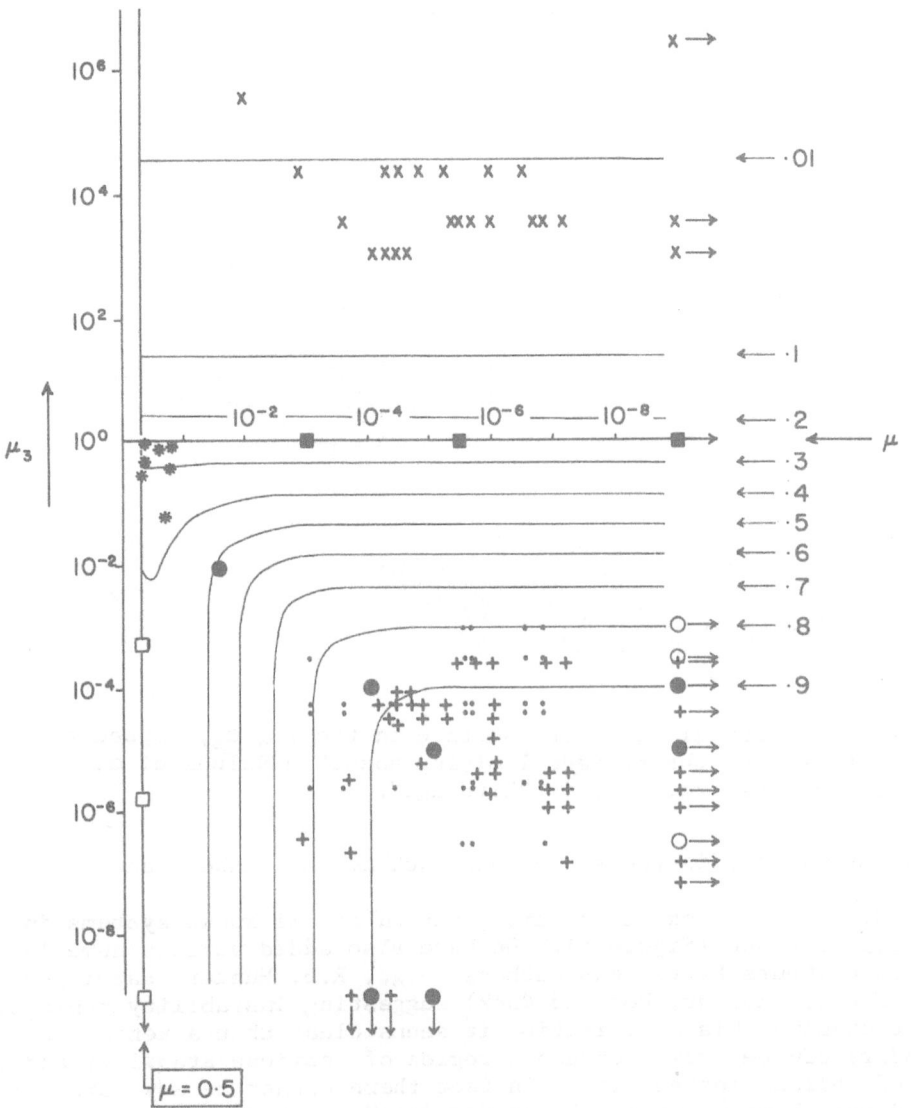

DISTRIBUTION OF KNOWN SYSTEMS IN μ, μ_3-PLANE

— AND CONTOURS OF CRITICAL STABILITY SURFACE

Figure 3. Distribution of known three-body systems in the μ,μ_3
plane and their relation to the contours of the critical
stability curface (see text for the meaning of the symbols).

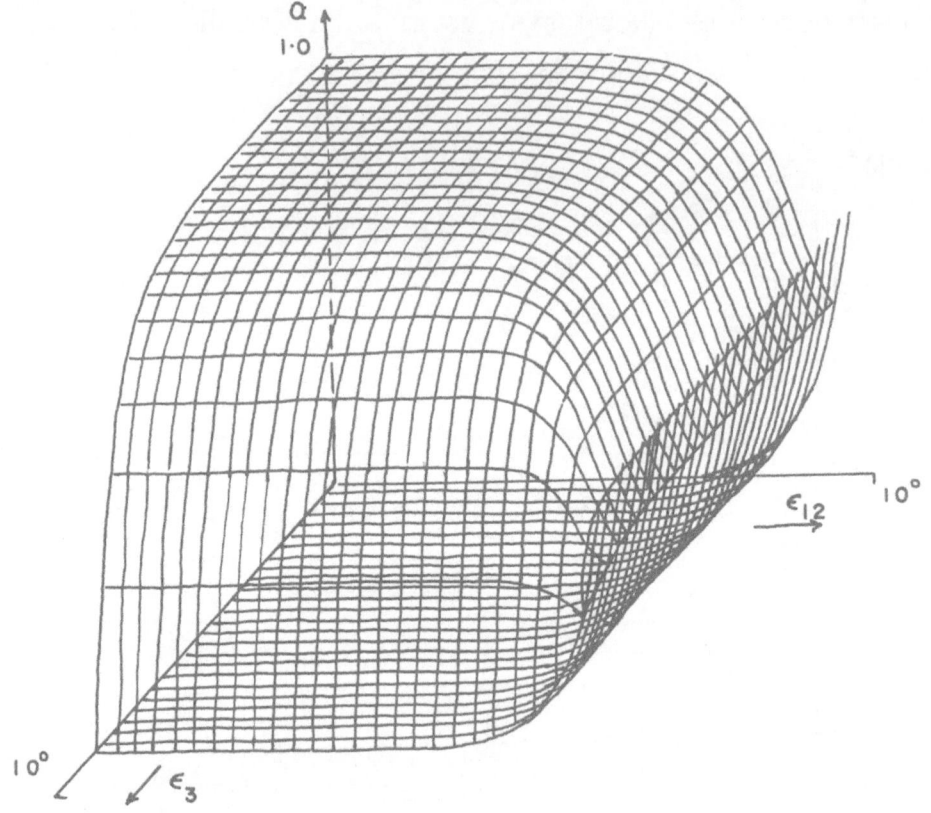

Figure 4.   Critical stability surface in the $\varepsilon_{12},\varepsilon_3,\alpha$ space together with the surface limiting possible values of α.   To-
gether they form Zare's "birdcage".

they may wander, in the sense that exchange may take place.

Now let us look at the distribution of all known systems in
the $\varepsilon_{12},\varepsilon_3$ plane (Figure 5).   We have also added various numerical experiments by various authors (e.g., R.B. Hunter, Harring-
ton, Nacozy, Horedt, Pop and Ruck) suggesting instability regions.
If we examine this distribution it seems clear that a tentative
boundary can be drawn between a region of obvious stability and
one of obvious instability.   In fact there cannot be any hard
boundary: perhaps it should be "smudged" out wider to indicate
a grey area of doubtful stability.

If we now project the birdcage's contours on to the $\varepsilon_{12},\varepsilon_3$
planet we obtain Figure 6.   The agreement is remarkable though,
by the nature of the definitions of ε_{12} and ε_3 care must be taken
in interpreting this felicitous situation.

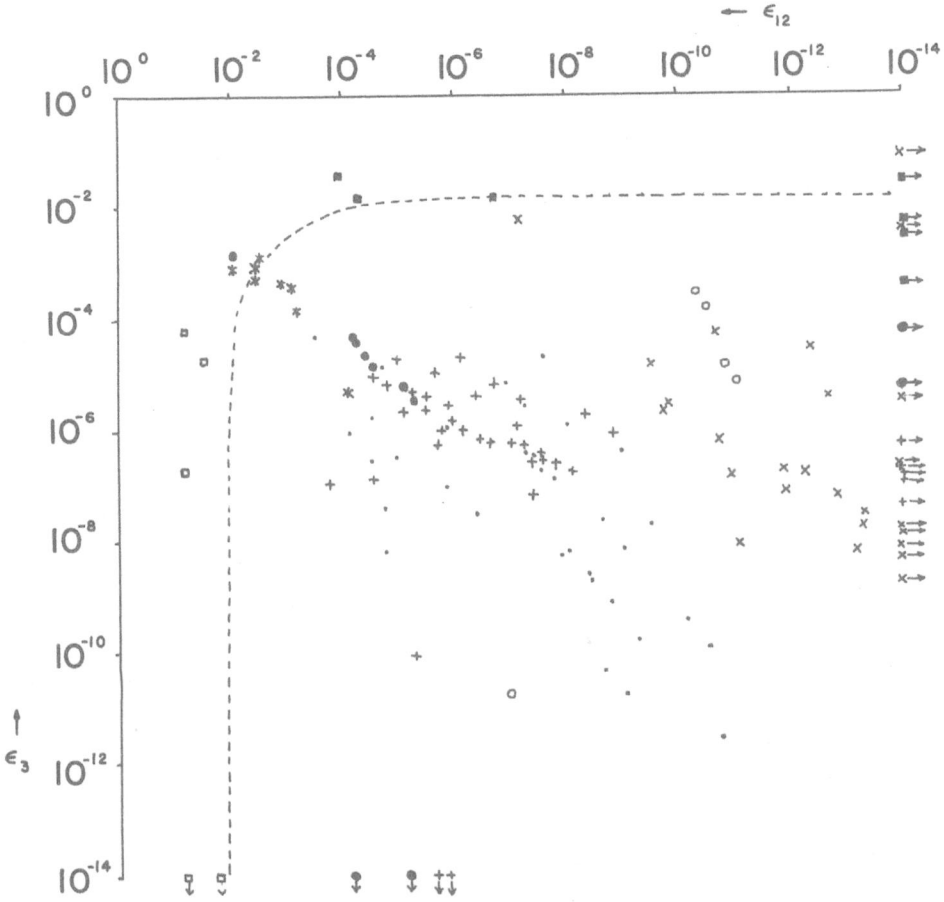

Figure 5. Distribution of known three-body systems in the $\varepsilon_{12}, \varepsilon_3$ plane. The dotted line separates empirically a region of stability $\varepsilon_{12} < 10^{-3}$, $\varepsilon_3 < 10^{-3}$ from a region of high instability $\varepsilon_{12} \gtrless 10^{-3}$, $\varepsilon_3 \gtrless 10^{-3}$. (See text for the meaning of the symbols)

For any given triple system we have three pieces of data, viz., μ, μ_3, α. But the first two also, by the critical stability criterion, dictate a value α_c of α. For any given real system defined in the $0\alpha, \mu, \mu_3$ coordinate system there is an α_{cr} for that system on the critical stability surface directly above or below the point denoting the real system.

Thus we have two sets of coordinates $(\alpha_{act}, \mu, \mu_3)$ referring to the real system, and $(\alpha_{cr}, \mu, \mu_3)$ referring to the critically stable system.

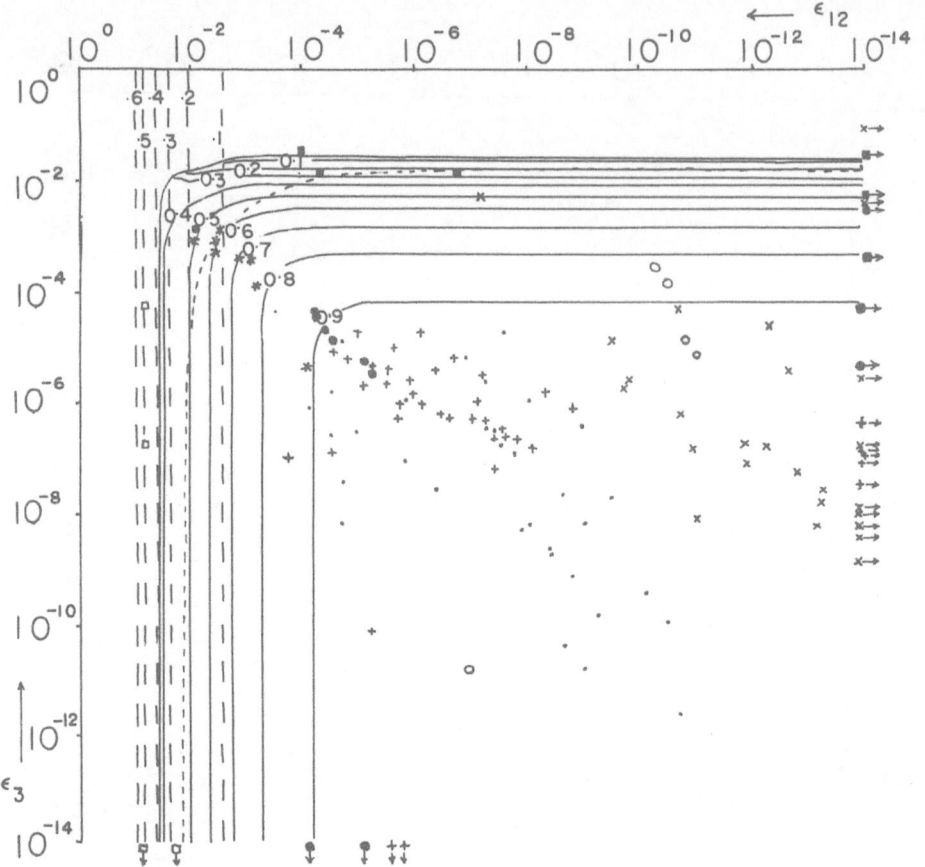

Figure 6.   Zare's "birdcage" contours of critical stability and
      contours limiting α-values projected onto the ϵ_{12}, ϵ_3 plane
      showing agreement with the division into regions of stability
      and instability.

        Clearly when transforming from the $0\alpha,\mu,\mu_3$ coordinate system
to the $0\alpha,\epsilon_{12},\epsilon_3$ coordinate system, the lines joining the two
points will no longer be perpendicular to the plane $\alpha = 0$.   The
relative shift of the two points can be given quantitatively as
follows:

        Let ϵ_{12},ϵ_3 be the position of the real system (α_{act},μ,μ_3) in
the ϵ_{12},ϵ_3 plane.   Let $\epsilon'_{12},\epsilon'_3$ be the position of the critically
stable system (α_{cr},μ,μ_3) in the ϵ_{12},ϵ_3 plane.

Then

$$\varepsilon_{12} = \mu(1-\mu)\alpha_{act}^2 \quad ; \quad \varepsilon_{12}' = \mu(1-\mu)\alpha_{cr}^2$$

$$\varepsilon_3 = \mu_3\alpha_{act}^3 \quad ; \quad \varepsilon_3' = \mu_3\alpha_{cr}^3$$

so that

$$\frac{\varepsilon_3}{\varepsilon_3'} = \left(\frac{\alpha_{act}}{\alpha_{cr}}\right)^3 \quad ; \quad \frac{\varepsilon_{12}}{\varepsilon_{12}'} = \left(\frac{\alpha_{act}}{\alpha_{cr}}\right)^2 .$$

There are three possible cases depending on whether the real system is (i) unstable, (ii) critically stable, or (iii) stable.

(i) In this case $\alpha_{act}/\alpha_{cr} > 1$ and $\varepsilon_{12} > \varepsilon_{12}'$; $\varepsilon_3 > \varepsilon_3'$, so that the unstable system is moved, in the $0\alpha,\varepsilon_{12},\varepsilon_3$ coordinate system to a position where the critical stability surface is lower than it was previously in the $0\alpha,\mu,\mu_3$ coordinate system.

(ii) Here $\alpha_{act}/\alpha_{cr} = 1$ and $\varepsilon_{12} = \varepsilon_{12}'$; $\varepsilon_3 = \varepsilon_3'$, that is, the critically stable system remains in the same position relative to the critical stability surface.

(iii) In this case $\alpha_{act}/\alpha_{cr} < 1$ and $\varepsilon_{12} = \varepsilon_{12}'$; $\varepsilon_3 < \varepsilon_3'$. Here the unstable system is moved to a position where the critical stability surface is higher.

Hence it can be seen that under no circumstances can a body change from one side of the critical stability surface to the other when the coordinate transformation is implemented. Note that the above three conditions are made more readily comprehensible due to the simplicity of the critical stability surface, i.e., it has no irregularities.

It can thus be considered that there may well be some sorting out of three-body systems depending on how "stable" or "unstable" they are. A large amount of numerical experimentation, integrating the equations of motion for various hypothetical three-body systems is required to confirm this conjecture. This sorting out would result in the two-dimensional representation in the $\varepsilon_{12},\varepsilon_3$ plane being sufficient for an indication of stability of any three-body system.

It is to be stressed that all the data presented here have been calculated for circular orbits, that is, each component of the binary moves in a circle about their common centre of mass and the third body moves in a circle about the centre of mass of

the two-body system. If eccentricities were taken into account
then the critical stability surface in both cases would be de-
pressed. We are at present carrying out work on this. At this
stage, however, we may draw certain conclusions.

8.   CONCLUSIONS

1. The surface of critical stability, even though it is
created for the coplanar general three-body problem, is relevant
to the three-dimensional general three-body problem, at least for
low inclinations, giving information about the stability of such
systems. Moreover, it is relevant to many-body systems, $n > 3$.
For such systems where there is an observable dynamical hierarchy
it surely can be said that for the many-body system to have any
chance of stability, its triple subsets should satisfy the three-
body surface of the critical stability criterion. By dynamical
hierarchy is meant the sort of systems observed in the solar
system, e.g., planets in orbit about the Sun, or satellites about
a planet, or a planet with a satellite disturbed by the sun or by
a stellar system consisting of a binary plus a third star and so
on: a triple subset should involve a binary and a third body.

It is worthwhile remarking that in a many-body system's dynamics, it is one of the features of classical perturbation theory
that in first-order theory, the perturbations on any body due to
the other bodies' actions are additive so that to that extent
a many-body system _is_ a set of triple systems. In higher-order
perturbations cross-effects come in, of course.

This epsilon representation of real systems and numerical
systems and their distribution with respect to the critical stability contours, seems to support the conjecture previously stated
that for an n-body system to be stable over an astronomically
long time, then

$$\sum_{j} |\varepsilon_{ij}| < \ell < 1, \qquad i = 2, 3, \ldots, n$$

where the ε's are particular functions of the relative radius
vectors and ℓ is about 10^{-3} to 10^{-4}.

2.   Although the analysis was carried out for zero eccentricity, we can draw attention to its implications for two remarkable systems. The first is Saturn-Titan-Hyperion. Relevant
data are $\mu = 1/4151$, $\mu_3 = 1/(5 \times 10^6)$, $\alpha = 0.8251$. For zero eccentricity, $\alpha_c = 0.86$. But the effect of the considerable eccentricity of Hyperion's orbit, namely 0.1, will decrease the value
of α_c. In addition, it might appear that the parameter α should

be more realistically given by

$$\alpha = \frac{\rho_{T_{max}}}{\rho_{H_{min}}} = \frac{\alpha_T(1+e_T)}{\alpha_H(1-e_H)} = 0.92.$$

The orbits are almost coplanar and so, by the critical stability criterion, the system should be unstable. But it exists. However, there is an additional factor to be taken into account, namely the critical argument θ resulting from the 3:4 commensurability in mean motions n and n' such that

$$\theta = 4\lambda - 3\lambda' - \tilde{\omega} \sim 180°; \quad 4n - 3n' - \dot{\tilde{\omega}} = 0,$$

λ and λ' being the longitudes of Hyperion and Titan, $\tilde{\omega}$ being the longitude of Hyperion's apse. The amplitude of θ is about 36°. This ensures that the satellite's conjunction line librates about the moving aposaturnium of Hyperion so that the effective α is never more than 0.78 which must be sufficiently below α_c to ensure stability. Colombo and Franklin (1973) have argued that even if Goldreich's tidal mechanism is not the cause of the Titan-Hyperion resonance it may have arisen naturally. In other words, it is possible that Titan and Hyperion were formed at that resonance and, because the system is stable, remained there. What we are saying in support of Colombo and Franklin's suggestion is that other pairs of satellites, formed as close together as Titan and Hyperion but without their resonance, would have dispersed because their α's were outside the critical value α_c.

The second system is Sun-Neptune-Pluto. Here, the data are:

Body	Semimajor axis A.U.	Eccentricity e	Mutual Inclination	Mean Motion (0/day)
Neptune	30.06	0.0096	} ~ 17°	0.005981
Pluto	39.44	0.2502		0.003979

It is well-known that Pluto's perihelion is nearer the Sun than Neptune's mean radius vector. Up until ten years ago, because of the possibility of close approaches of Pluto to Neptune and the possibility of $\alpha = \rho_N/\rho_P > 1$ when Pluto was at perihelion, standard expansion techniques such as the one sketched earlier in the Jacobian equations of motion could not be applied to the Pluto-Neptune system because of the non-convergence of series involving $(\rho_N/\rho_P)^i$. In the context of the present study, we would say that the critical stability criterion for Neptune-Pluto would show that the system was unstable, since α can be greater than α_c and indeed greater than unity.

It is now known of course from the work of Cohen, Hubbard and Oesterwinter, that a critical argument $\theta = 3\lambda_P - 2\lambda_N - \tilde{\omega}_P \sim$ 180 exists which librates about 180° with an amplitude of 76° and period 19500 years, thus ensuring that conjunctions of the two planets avoid Pluto's perihelion. The maximum effective α is thus never more than 0.75. It is also possible that the argument of Pluto's perihelion also librates, a suggestion put forward by Brouwer (1966) who pointed out that a critical argument $\theta' = 3\lambda_P - 2\lambda_N - \Omega_P$ should exist ensuring that $\omega = \theta' - \theta$ would librate. If so, then the high inclination of Pluto's orbit to Neptune's

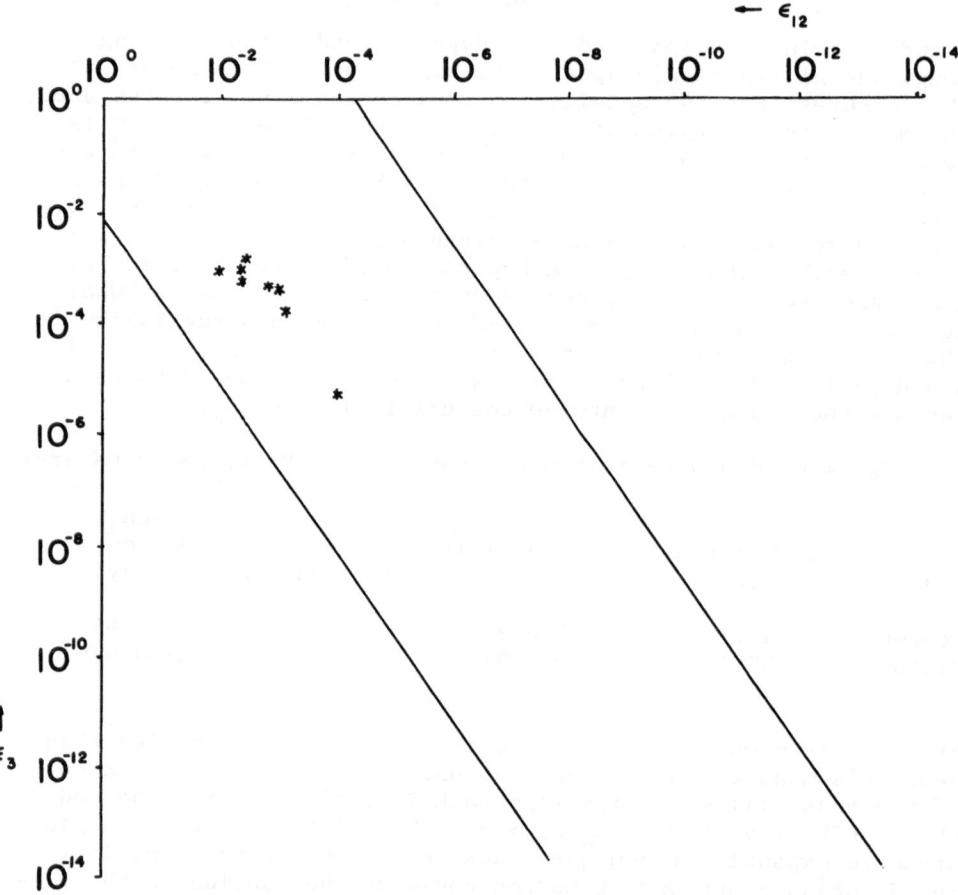

Figure 7. Plot of eight data for eight triple stellar systems. The two diagonal straight lines enclose a region within which stellar three-body systems might be expected to be found on mass distribution data alone.

could also increase the stability. Williams and Benson's study of 1971 showed that indeed ω librated about 90° with an amplitude of approximately 24°.

3. Finally, let us consider the distribution of certain real cases of triple systems on the $\varepsilon_{12}, \varepsilon_3$ plane together with the critical stability contours.

Our findings are that the bottom right hand region bounded by $\varepsilon_{12} = \varepsilon_3 = 10^{-3}$ is a region of high stability in the many body problem.

In Figure 7 we plot the available data for triple star systems, the two diagonal lines denoting an area within which triple star systems might be expected to exist according to the observed range of values of star masses and according to the range taken for α, viz., 0.85 to 10^{-4}. The high end of the α-range would obviously provide highly unstable triples in many cases. The eight asterisks denote the posibitions of the eight triple star systems given in Szebehely and Zare (1977).

Apart from the fact that several of these systems are only marginally stable, the outstanding feature of Figure 7 is the complete lack of data for the region of maximum stability. There are three possible explanations:

1. We have not plotted all the available data.
2. There is an observational selection effect making it easier to detect triples in the upper left hand section.
3. They do not exist.

With respect to the first reason it is certain that scattered throughout the literature lie additional data for triple stellar systems: a future search of the literature will harvest more points, possibly in the stable region.

There could well be an observational selection effect. However, for stellar systems, shifts in the ε-plane are predominantly α-shifts. The stable region for triple stars is one of low α-values and therefore high periods of revolution for the third star about the binary, unless the binary is a very close one. For most visual binaries that have a triple companion, therefore, two centuries of observation may well be inadequate to detect the very large number of triples that may exist.

The third explanation - their nonexistence - implies that for some reason, triples of this kind cannot exist. This is undoubtedly the case for very widespread binaries themselves, let alone triples, since the central mass of the galaxy itself plays the part of a third perturbing body rendering any binary star system unstable if the components are more widely separated than a certain limiting value. For data take two solar masses whose separation r is α times the distance R of the binary from the centre of the Galaxy, taken to be 10^4 parsecs. The Galaxy's mass may be taken to be 10^{11} solar masses.

Now we have

$$\varepsilon_3 = \frac{m_3}{m_1+m_2} \alpha^3 = \mu_3\alpha^3 \, ,$$

where $\mu_3 \simeq 10^{11}$. For stability we should take ε as 10^{-3} or smaller.

Then

$$\alpha = [\frac{\varepsilon_3}{\mu_3}]^{1/3} \simeq 1/50,000.$$

Hence r for limiting stability of binaries in the region of the sun is of order 0.2 parsecs, or 40,000 A.U. If for example we are looking for a triple stellar system whose third body is of order 40,000 A.U. from its binary and we wanted it to be in the region $\varepsilon_{12} \simeq 10^{-6}$, $\varepsilon_3 \simeq 10^{-9}$, we would have $\alpha \simeq 10^{-4}$ with $\varepsilon_{12} \simeq 10^{-8}$ and $\varepsilon_3 \simeq 10^{-12}$. There is no way in which such systems could have been detected.

We now look at the planets of double stars.

The points in Figure 8 are given by numerical experiments by Harrington (1977) delineating very roughly the boundary of the intrinsically unstable region for high $\varepsilon_{12},\varepsilon_3$, i.e., $\varepsilon > 10^{-3}$. The lines denote the boundaries of the regions within which such systems could exist merely by using reasonable ranges for the masses of planets and stars. A considerable part of each region lies within the stable region of the ε-plane where $\varepsilon \leq 10^{-3}$.

Now the detection of planetary systems of stars other than the sun is impossible except for the very nearest stars. It is not surprising then that no points have been plotted in the stable region. Some of the remarks made about the existence of triple stellar systems in this stable region may be applied to hypothetical planets of binary stars. For example, if we are dealing with inferior planets with the companion of the binary perturbing them, the limiting separation of both stars imposed by the presence of the galactic central mass is again 40,000 A.U., at least in the sun's neighborhood. On the other hand, for a superior planet of a binary star system, the planet is the third body and cannot exist further than 40,000 A.U. from its binary.

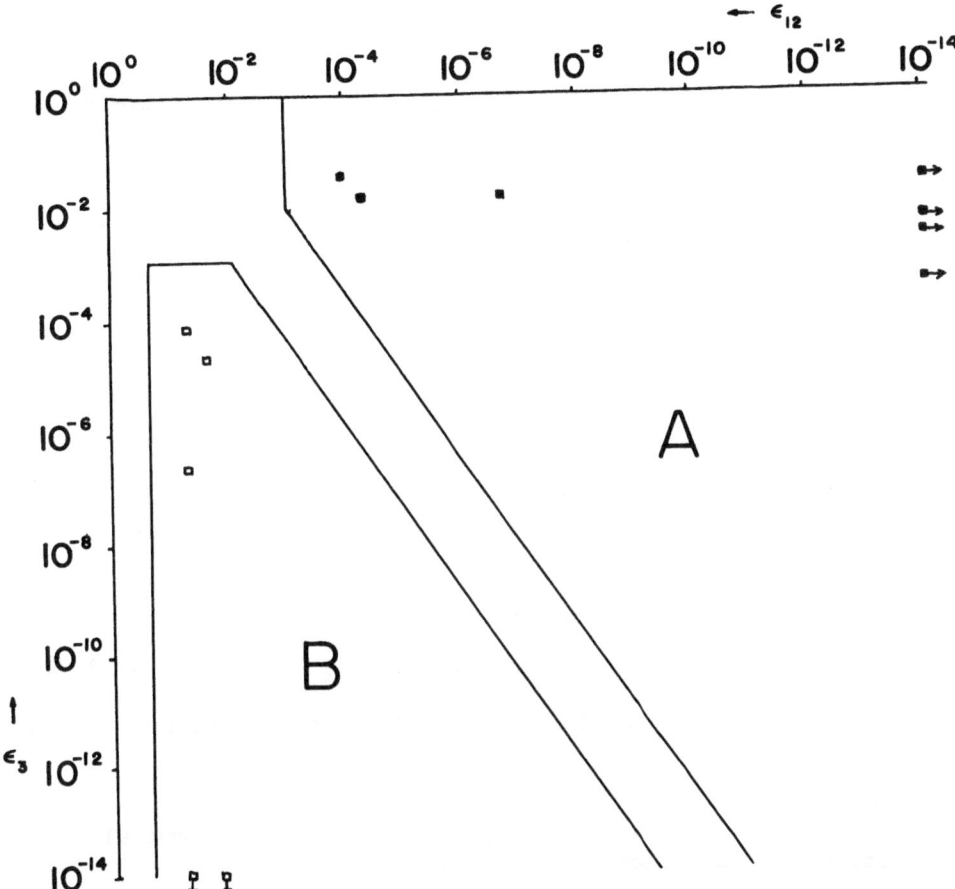

Figure 8. Superior and inferior planets in double sun systems
from computations by Harrington where the points plotted
roughly delineate the ε_{12} and ε_3 values at which instability
occurs. The straight lines enclose regions within which
such planets might be expected to be found solely on mass
distribution data of stars and planets.

Unlike the triple stellar case, the small mass of the planet
produces, for an inferior planet, a low ε_{12} and a high ε_3, giving
points in the stable region while a superior planet gives a high
ε_{12} and a low ε_3 giving points in the stable region. As far as
the ε-diagram and the perturbation by the galaxy is concerned,
therefore, there is no reason why planetary systems of binary
stars should not exist.

The work by Roy, Emslie and Walker summarized in these lectures will be the subject of a more detailed paper in course of
preparation.

9. REFERENCES

Brouwer, D. (1966), in "The Theory of Orbits in the Solar System and in Stellar Systems" (ed. G. Contopoulos), Academic Press, N.Y., p. 227.

Cohen, C.J. and E.C. Hubbard (1965), Astron. Journ. 70, 10.

Cohen, C.J., E.C. Hubbard, and C. Oesterwinter (1967), Astron. Journ. 72, p. 973.

Cohen, C.J., E.C. Hubbard, and C. Oesterwinter (1972), Astron. Pap. Amer. Ephemeris, 13.

Cohen, C.J., E.C. Hubbard, and C. Oesterwinter (1973), Cel.Mech. 7, p. 438.

Colombo, G. and F.A. Franklin (1973), in "Recent Advances in Dynamical Astronomy" (ed. B. Tapley and V. Szebehely), Reidel Publ. Co., Holland.

Harrington, R.S. (1977), Astron. Journ., 82, p. 753.

Horedt, G.P., P. Pop, and H. Ruck (1977), Cel. Mech. 16, p. 209.

Hunter, R.B. (1967), Mon. Not. Roy. Astron. Soc., 136, pp. 245,267.

Nacozy, P. (1977), Cel. Mech. 16, p. 77.

Plummer, H.C. (1918), "An Introductory Treatise on Dynamical Astronomy," Cambridge University Press.

Szebehely, V. (1977), Cel. Mech., 15, p. 107.

Szebehely, V. and K. Zare (1977), Astron. Astrophys., 58, p. 145.

Szebehely, V. and R. McKenzie (1977a), Astron. Journ. 82, p. 79.

Szebehely, V. and R. McKenzie (1977b), Astron. Journ. 82, p 303.

Williams, J.G. and G.S. Benson (1971), Astron. Journ. 76, p. 167

Zare, K. (1976), Cel. Mch., 14, p. 73.

Zare, K. (1977), Cel. Mch., 16, p. 35.

INSTABILITIES IN PLANETARY SYSTEMS

R. O. Vicente

Department of Applied Mathematics, Faculty of Sciences,
Lisbon University, Portugal

The consideration of stability and instability when applied
to planetary systems depends not only on the definitions of these
terms but also the intervals of time we are envisaging. Another
question refers to what sort of planetary systems we are thinking about.

Let us first consider the subject of the dynamics of planetary systems as opposed to stellar systems. We know an enor-
mous variety of stellar systems because stars can be observed
far more easily than planets. We also know that binary stars
are numerous in the universe, and we are not going to consider
multiple stars. We restrict ourselves to possible planetary
systems around single or binary stars.

For well known reasons, the only planetary system so far
observed is the one moving around the Sun. Considering the number of single stars in our galaxy, we can immediately ask the
question if any other planetary systems will present the same
dynamical features as our own. We have, therefore, a great lack
of observational knowledge about the composition, structure and
dynamical motions of planetary systems. Let us try to forecast
the possible structure of planetary systems because if we have
a good enough theory we might guess the future. It is well
known that futurology, the science of forecasting the future,
is a very hazardous science, but it is worthwhile and a good
exercise to see what might be discovered in the future.

I should like to give you the example of Lagrange's three
particles, discovered by Lagrange in 1772, where there are an
infinite number of solutions of the problem of three bodies, in

Victor G. Szebehely (ed.), Instabilities in Dynamical Systems. 211-225.
Copyright © 1979 by D. Reidel Publishing Company.

which the triangle formed by the bodies remains equilateral and
of constant sizes and rotates uniformly in the plane of the motion.
This discovery was considered, at the time, to be of theoretical
interest only and no one tried to observe particles in the solar
system, satisfying these conditions. That means astronomers did
not think it worthwhile to spend time observing the celestial
sphere in search for such bodies because they probably thought
the theory was not good enough, and we had to wait until 1906,
when the minor planet Achilles was observed to have a mean distance equal to that of Jupiter. It was the first time that an
approximate example of the Lagrangian equilateral-triangular
configuration was detected, and it took nearly 150 years to observe the group of minor planets later called the Trojan group.

A question I should like to ask is whether our present day
theories are good enough to forecast other types of planetary
systems besides our own, the only one so far, accessible to observation. It would be interesting if we could forecast other
and more complex types of planetary systems, so that our colleagues in the next few centuries would already have an idea
about them. Related to this question, appears immediately, the
problem of stable and unstable planetary systems. We can imagine
planetary systems where the primaries have widely different mass
ratios, and the number of satellites also has greater variations.
Another interesting possibility refers to the inclination and
eccentricities of the orbits. How complex can such a system be?

Let us now consider another class of planetary systems
formed around binary stars. This type of system should be the
second in abundance, considering the large number of known binary stars. It is possible to imagine a model where we have
planets moving around one or the other of the stars, where the
masses are quite different and where their attached satellites
are very numerous or not so numerous. To make the problem still
more difficult we can consider the planets having various inclinations and eccentricities for their orbits. Is it worthwhile to build such planetary models around binary stars?

Another aspect of the question of constructing models of
planetary systems refers to the consideration of the intervals
of time we are going to imagine for the evolution of the system.
We can consider the problem since the formation of the system
or applying only for a certain period of its evolution.

Let us consider the problem of time applied since the formation of planetary systems. We have already spoken about the
question of the possible number of planetary systems and there
is no doubt that there will be an appreciable number of such
systems but, from the more strict scientific point of view, we
should say that there is a high "a priori" probability

(Jeffreys, 1967) for the existence of such planetary systems.
We know, so far, only one such system but considering present
day ideas about the formation of planetary systems we can assign
a high probability to this event.

Considering, therefore, there is an appreciable number of
planetary systems we may ask other questions referring to the
time of its formation. Were all of them formed at the same time?
Or are they being formed all the time? The answers to these two
questions depend on the theories adopted for the formation and
evolution of the universe. We are not concerned here with such
problems and, for our purposes, we can adopt the hypothesis that
they were formed, at least some of them, since the beginning of
the universe.

We have to consider now the problem of the age of universe.
The different hypotheses about the age of the universe have
varied since astronomers started thinking about this cosmological
problem. Laplace (1830), about 150 years ago, wrote (Vol. 1,
pp. 1,2):

"Of all the natural sciences, astronomy is that which
presents the longest series of discoveries. The first
appearance of the heavens is indeed far removed from
that enlarged view, by which we comprehend at the
present day, the past and future states of the system
of the world. To arrive at this, it was necessary to
observe the heavenly bodies during a long succession
of ages, to recognize in their appearances the real
motion of the earth, to develope the laws of the
planetary motions, to derive from these laws the
principle of universal gravitation, and finally from
this principle to descend to the complete explanation
of all the celestial phenomena in their minutest details. This is what the human understanding has
achieved in astronomy. The exposition of these discoveries, and of the most simple manner, in which
they may arise one from the other, will have the two-
fold advantage of furnishing a great assemblage of
important truths, and of pointing out the true method
which should be followed in investigating the laws
of nature."

Since this first tentative explanation of the system of
the universe, different theories have systematically increased
the age attributed to the universe since its origins, if there
was an origin for the universe.

We can consider a value of 4500 million years for the age
of the solar system, in agreement with present day ideas about

this problem, but we must be aware of past mistakes done on this
subject, and one of the main difficulties is lack of adequate
observations and insufficient theoretical developments. Following a correct statistical procedure, we should assign a measure
of precision to the value mentioned above for the age of the
solar system. We can always adopt a more or less optimistic
view about the precision attached to this value, for instance,
ten or twenty five per cent, and considering a conservative
figure we might say that 4500 ± 1000 million years will comprise
the true value of the age of the solar system, or, we can be
more optimistic, and consider the following value 4500 ± 500
million years. Of course, even this figure has only meaning in
relation to present day models of the origin and evolution of
the universe. If the models adopted in actual researches are
too simplified or do not consider any physical phenomena not
yet discovered, and relevant to the problem of the age of the
universe, the figures quoted above might be completely wrong.

I should like to recall the discussions between Lord Kelvin
and the geologists, about a century ago, precisely on the age of
the sun and the solar system. Lord Kelvin says (Thomson and
Tait, 1883, p. 494):

"...the considerations deduced above, in this paper,
regarding the sun's possible specific heat, rate of
cooling, and superficial temperature, render it probable that he must have been very sensibly warmer
one million years ago than now; and, consequently,
that if he has existed as a luminary for ten or
twenty million years, he must have radiated away
considerably more than ten or twenty million times
the present yearly amount of loss.

It seems, therefore, on the whole most probable that the
Sun has not illuminated the Earth for 10^8 years, and almost
certain that he has not done so for 5×10^8 years. As for the
future, we may say, with equal certainty, that inhabitants of
the Earth cannot continue to enjoy the light and heat essential
to their life, for many million years longer, unless sources
now unknown to us are prepared in the great storehouse of
Creation."

This is a good example of the history of science about
the shortcomings of theoretical mathematical models, and should
make us aware about relying too much on present day knowledge.
So, we can accept the figure above mentioned, if we consider the
premises stated that there are no unknown variables that might
substantially modify the physical mathematical models so far
constructed.

We have now an interval of time that will be employed for our definitions of stability and instability in planetary systems. We can deal with this problem in different ways, and we can consider the stability of the system since the origin or for certain intervals of time. Of course, the intervals of time can be measured in units of 10^6 years or for 10^9 years, because it might happen that the system will be stable for a period of 10^6 years but unstable for a longer period of time.

The statement that planetary systems have been stable since their formation does not correspond to a valid statement, because we can verify that not only from the behaviour of theoretical models but also from the observations showing several results of instability in the solar system. Of course, following a proper scientific thinking we might say that our solar system shows such instabilities but other systems might exist where such instabilities do not apply.

It is probably convenient to classify the span of time since the formation of the solar system in the following intervals:

> up to 10^3 years - very short
> up to 10^5 years - short
> up to 10^7 years - long
> up to 10^9 years - very long

Considering this simple classification we see immediately one of the great difficulties we have when dealing with the solar system. All our astronomical observations correspond only to periods of time that we consider as very short. One of the longer records of systematic astronomical observations correspond to the luni-solar precession, observed since about 2,000 years ago, and that is one of the longest periods; following our classification, would be a very short interval of observation.

The problem that appears is the validity of such observations for the longer periods of time, that is, if it is a correct scientific procedure to extrapolate such results of the observations for longer time intervals, for instance, 10^7 years.

We have still a shorter interval in the case of the observations of the motions of the planets which have been done in a more reliable and regular way only in the last 200 years, that is, about ten percent of the time during which we have observed the luni-solar precession.

Laplace (1830, 2, 325) mentions:

"Astronomy thus becomes the solution of a great

problem of mechanics, the constant arbitraries of
which are the elements of the heavenly motions. It
has all the certainty which can result from the im-
mense number and variety of phenomena, which it rigorously explains, and from the simplicity of the
principle which serves to explain them. Far from
being apprehensive that the discovery of a new star
will falsify this principle, we may be antecedently
certain that its motion will be conformable to it;
indeed this is what we ourselves have experienced
with respect to Uranus and the four telescopic
planets recently discovered, and every new comet
which appears, furnishes us with an additional proof."

We can agree to a certain extent with these statements, but
the words "rigorously explains" depend not only on the precision
of the observations but also the span of time that applies to the
motions of the system. We can think about the stability of
planetary systems for long and very long periods of time, and the
problem becomes more difficult.

Our present dynamical theories of the motions of the solar
system can explain the motions for a few centuries, and we have
not checked them for more than 3 centuries. Even this statement
needs to be qualified in regard to the precision of the observations, because the precision obtained with the instruments of
the 17th century is far lower than present day precision. The
question still remains to know if our dynamical theories will
be verified for the next few centuries with 20th century or higher
precision; considering our classification of time intervals, we can
mention 10^3 years, and the problem becomes more difficult. If we
find small discrepancies between theory and observations, we can
assign them to the theory, saying that it is imperfect, or to as-
cribe them to an instability of the system which we are observing.

This raises the question of the stability and instability
of the solar system for the time intervals mentioned. Let us
try to think from the beginning of the formation of the solar
system, corresponding to a period of great instability, if we
really can classify instabilities in order of magnitude. It is
possible to say it was a time of great instability in comparison
with periods in the evolution of the solar system, when the instability did not manifest itself in such shorter periods, or
did not apply to so many bodies that constitute the system.

Classifying, then, that stage of the evolution as corresponding to a great instability, we have to assign an interval of
time for the duration of the instability. This is a difficult
question, depending on the theories adopted for the formation of
the solar system, but let us try a few guesses. If we admit that

the instability lasted a very short period (up to 10^3 years), this corresponds to a ratio of one part to the 10^6 of the duration of the system, or if we admit a short period (10^5 years) we have a ratio of one to 10^4. These figures might give us a lower and upper bound for this stage of instability, and might satisfy any school of cosmologists, not only the ones that think the instability lasted a very short period, but also the others thinking that it took longer for the solar system to settle into a stable configuration. We should also pay attention, and define, what is meant by the end of this stage of evolution. It is a reasonable assumption to think that it corresponds to the existence of well defined orbits, at least, for the planets.

We enter now a stage in the evolution of the solar system which is considered as stable, and we have again to classify that stability. How long did this stability last? Could it be stable for the planets and unstable for the satellites? Or could we even have instability in the orbits of some planets? And what about the asteroids?

It would be interesting to answer these questions with a reasonable degree of certainty. We could consider the overall configuration of the system as stable, but some regions of it showing different degress of instability.

Quoting Laplace (1830, 2, 327-9):

"Another phenomenon of the solar system equally remarkable, is the small excentricity of the orbits of
the planets and their satellites, while those of comets
are very much extended. The orbits of this system
present no intermediate shades between a great and
small excentricity. We are here again compelled to
acknowledge the effect of a regular cause; chance
alone could not have given a form nearly circular
to the orbits of all the planets. It is therefore
necessary that the cause which determined the motions
of these bodies, rendered them also nearly circular.
This cause then must also have influenced the great
excentricity of the orbits of comets, and their motion in every direction;....

What is this primitive cause? In the concluding note
of this work I will suggest an hypothesis which appears
to me to result with a great degree of probability,
from the preceding phenomena, which however I present
with that diffidence, which ought always to attach
to whatever is not the result of observation and
computation.

Whatever be the true cause, it is certain that the
elements of the planetary system are so arranged
as to enjoy the greatest possible stability, unless
it is deranged by the intervention of foreign causes.
From the sole circumstance that the motions of the
planets and satellites are performed in orbits nearly circular, in the same direction, and in planes
which are inconsiderably inclined to each other, the
system will always oscillate about a mean state, from
which it will deviate but by very small quantities.
The mean motions of rotation and of revolution of
these different bodies are uniform, and their mean
distances from the foci of the principal forces which
actuate them are constant; all the secular inequalities are periodic. The most considerable are those
which affect the motions of the Moon, with respect
to its perigee, to its nodes and the Sun; they amount
to several circumferences, but after a great number
of centuries they are reestablished."

The last paragraph of this quotation mentions two subjects
that interest us. The first subject refers to the great stability of the system, adding very wisely "unless it is deranged by
the intervention of foreign causes". But even if we do not consider foreign causes, we have to consider the stability of the
system in relation to time, and that is the important problem.

The second subject refers to the secular inequalities that
are periodic. Laplace mentions their periodicity after a great
number of centuries, but does not specify how many centuries.
We have again the same problem of the periodicity of these inequalities for different spans of time. Are they valid for very
short and short periods? Or can we guarantee that the theory
is so good that they are valid for long periods up to 10^7 years?
We can also ask the question if there is any meaning in speaking about stability for very long periods, that is, to the order
of 10^9 years. Again, as we mentioned before, the checking of
the theory can only be done through observations which are necessarily correlated with the precision of our instruments.

A great difficulty we have is the very short time when our
observations have been made, and we cannot observe other planetary systems at different stages of their evolution, as we are
able to do with stellar systems. It is therefore not possible
to check the stability of the solar system within such a short
interval of our observations. We can raise the question if it is
a correct scientific procedure to extrapolate the results, obtained
by mathematical models that we are hopelessly incapable of checking with our inadequate set of observations.

The word stability appears already in the work of Laplace, but we are nowadays able to give it a mathematical formulation. There are different points of view that have been employed to define stability. We can consider Liapunov's stability, connected with the concept of isochronous correspondence, while orbital stability is related to normal correspondence.

We can also consider the stability of equilibrium solutions, and if a dynamical system is in a state of equilibrium, it remains in that state as $t \longrightarrow \infty$. This corresponds to a pure mathematical model which might be, more or less, related to the actual dynamics of the solar system.

Another way of looking at the stability of a dynamical system corresponds to relating its behaviour to disturbances that affect the system in the course of time. This is a problem that can be applied to planetary systems, and we can consider the disturbances as being periodic or increasing continuously with time; also we have to consider what intervals of time we are concerned with: very short, short, long and very long. This offers a wide range of possibilities with practical interest.

It has been considered that the motion which remains in the small neighbourhood of the equilibrium point, after it is disturbed, is called stable. We have, therefore, another way of looking at stability, and we can give it a mathematical formulation (Szebehely, 1967).

The definitions of stability and instability, in a topological space, are given by Siegel (1956), and this approach corresponds to the point of view of pure mathematics.

The mathematical formulation of the several definitions of stability are useful to give us ideas about the possible behaviour of planetary systems, but we must not forget they correspond to mathematical models that do not take account of the physical complexities of the system.

One of the important problems of dynamics is to find the effect of the perturbations on the behaviour of orbits for very long intervals of time, and which is expressed mathematically as $t \longrightarrow \infty$.

We can mean by perturbation, the deviation of a dynamical system from an integrable system, but the problem of integrability is a very delicate problem of mathematical analysis. From the point of view of observational astronomy, any observations are affected by errors, and the motions of any dynamical system are known with a certain precision. This precision has increased in the course of the last centuries, but it may not be sufficient

to distinguish between integrable and non-integrable systems.
We cannot avoid this question, and we can only hope that the precision of our observations will improve in the future.

Quoting Laplace (1830, 2, 338-9):

"Astronomy has already made an important step, in
making us acquainted with the motion of the Earth,
and the epicycles which the Moon and the satellites
describe on the orbits of their respective primary
planets. But if ages were necessary in order to
know the motions of the planetary system, what a
great length of time must be required for the determination of the motions of the Sun and the stars;
notwithstanding this, such motions appear to be
already indicated by observations.

The progress of astronomy depends on these three
things: the measure of time, that of angles, and
the perfection of optical instruments. The two
first are nearly as perfect as we could wish; it
is therefore to the improvement of the latter that
our attention should be directed,..."

Laplace's three things for the progress of astronomy are
still actual but his statement that the measures of time and
angles do not need any improvement do not correspond to his far-
reaching views on the evolution of the world, showing that,
sometimes, he did not employ a correct scientific language.

Our present day time-keeping devices, the atomic clocks,
show how great the improvement has been in the last 150 years;
unfortunately, the improvement in the measurement of angles is
not so good. We wonder what will happen to the three things
mentioned by Laplace, in the next 150 years!

In the problems of stability, we can consider a case that
more nearly approaches the motions of the natural satellites,
and has been dealt with by Szebehely (1978). This paper gives
an interesting application to the solar system, but stating
clearly the restrictions of the mathematical model employed. The
conclusions obtained, that all known natural satellites are inside the stability limit, except the four outermost satellites
of Jupiter, give us a hint about possible lines of evolution for
the satellites.

Considering this model, and looking at Fig. 2 of the paper,
we might give a possible interpretation of some of the main features of the natural satellites:

NEPTUNE

Thinking about the features of the elements of the orbits of the two satellites, we notice that Neptune presents an interesting case of the evolution of the satellites. These features are: 1) Nereid has one of the greatest eccentricities of the satellites of the solar system, an appreciable inclination of the orbit (27°) in relation to the planet's equator, and the great difference of its inclination from Triton's inclination; 2) the motion of Triton is retrograde and its orbital plane is inclined 20° to the equatorial plane of Neptune.

It is probably a good guess to think that the two satellites have passed through different stages of stability and instability, because of such different features of their orbits.

Looking at Fig. 2, we see that Nereid is nearer to the instability region than Triton. If we could consider a more complicated mathematical model, we might be able to offer a better explanation of these features. We can also imagine some stability problems for Nereid, applying for a long interval of time, for instance, 10^7 years.

URANUS

The orbits of the five satellites have very small eccentricities and lie in the same plane, according to our present day observations; the common orbital plane is presumably the plane of the equator of the planet.

Looking at Fig. 2, we see that Miranda and Oberon are well within the stability region. It looks like this mathematical model is satisfactory for the satellites of Uranus, considering the similar features of all of them. We might say that they are stable at the moment, but an interesting question to ask would be the following: how long has this stability lasted?

SATURN

The two satellites with greater eccentricity of their mean orbits are Hyperion and Phoebe, while Iapetus and Phoebe have the greatest inclination of their orbits in relation to Saturn's equator. The motion of Phoebe is retrograde.

We notice that satellites I (Mimas) to VII (Hyperion) have the inclinations of their orbits within 1°.5 of the planet's equator, and two VIII (Iapetus) and IX (Phoebe) having appreciable inclinations. What is the meaning of these differences from the point of view of stability?

Looking at Fig. 2, we see that Mimas is well within the stability region, but Phoebe is below the line of stability, showing a small margin of stability.

JUPITER

We can consider the satellites divided in three groups: 1) formed by the four Galilean satellites and Jupiter V, practically situated in the planet's equator and having nearly circular orbits; 2) Jupiter VI, VII and X showing inclinations of their orbits between 25° and 29° in relation to the planet's equator, and eccentricities of mean orbits from 0.13 to 0.21; 3) formed by satellites VIII, IX, XI and XII having retrograde motions and eccentricities of their mean orbits greater than any of the other groups.

Looking at the features of these three groups of satellites, we are tempted to infer some conclusions, saying that the first group shows stability, while the second group presents some possible instabilities and, finally, the third group showing definite signs of instability. These conclusions, obtained by simply examining the values mentioned, need to be taken with great care if they are not supported by any adequate mathematical model.

Actually, looking at Fig. 2, we notice that Jupiter V (first group) is well within the region of stability, while Jupiter VII (second group) is nearer to the line of instability, and the four retrograde satellites (third group) are situated in the instability region. This mathematical model seems to give us some possible explanations for the present day inclinations and eccentricities of Jupiter's satellites.

Referring the question of instability to the time intervals we have described, we might ask if the first group of satellites has been stable since the initial period of instability of the system. We might ask if the second group has entered a possible period of instability, how long has it lasted and how long will it last? Finally, the third group seems to be in a period of instability; will it last 10^5 or 10^7 years, before entering another period of greater instability or stability? We can also ask if it is possible to have a mathematical model that might give us some idea about the epoch when this instability appeared.

MARS

Phobos and Deimos are practically situated in the planet's equator, and have very small eccentricities of their mean orbits. Following a similar line of thinking, we might infer that they should be stable.

Looking at Fig. 2, this way of thinking seems to be corroborated because they are represented well within the region of stability.

We have, nowadays, close-up photographs of these two satellites, showing their dimensions and shape. Looking at the photographs, we might wonder how they have been formed and how their orbits evolved in order to show their present day values. Have they been captures? Can we assign any intervals of time for their present day stability?

EARTH

The case of our natural satellite has always been a special one, considering not only the features of its orbit but also the mass. Following all appearances, it seems an interesting case of stability.

Looking at Fig. 2, we see that the Moon is just below the line of stability, as it also happens with Phoebe and Jupiter VII. In spite of the mathematical model employed corresponding to the restricted problem, not taking into account the masses of the satellites and the eccentricities of the primaries, it gives us some indications about their stabilities.

This marginal stability of the Moon becomes a possible marginal instability when the dynamical model of the general problem of three bodies is employed, and the Moon might escape the Earth or it might be captured by it. The effect of the eccentricity of the Earth's orbit is pronounced on the measure of the stability of the system (Szebehely and McKenzie, 1977).

We have here a good example of how the modification from a restricted to a general problem of three bodies, changes the outlook about the possible evolution of the Earth-Moon system. This gives us reasons to wonder how the system would change for long intervals of time, and more general mathematical models.

****

The study of the stability of the planets offers us few

possibilities, because we only know of nine planets. It is,
therefore, difficult to find any trends of instability, all of
them moving in the same direction around the Sun. The planets
with larger inclinations of their orbits are Mercury and Pluto,
and they are also the planets with greater eccentricities of
their orbits. They correspond to the planets nearer to and further away from the Sun. Have these morphological appearances
any meaning for their stability in a long interval of time? How
long has this stage lasted in their evolution?

The study of the satellites offers more possibilities
because we have, at least, 32 natural satellites and some features
that have been considered strange in the past, might be explained
by the researches on the stability of satellites. We know that
some special conditions appear in a more or less long interval
of time, that lead to the actual behaviour of the system we observe.

The study of the asteroids offers an interesting opportunity
of studying possible cases of instability, because they are very
numerous. A statistical estimate suggests that about 30,000 are
within reach of modern instruments, but only five percent have
their orbital elements known with some precision. The majority
of orbits have a small inclination to the ecliptic and a small
eccentricity. It is known that these orbital elements do not
exhibit a random distribution, and there are correlations among
them. Besides that there are a number of orbits that show interesting and exceptional features, and the number of such orbits
will increase, if there is sufficient interest in the observations
of asteroids. Unfortunately, there are not many or regular pro-
grammes of observation of the asteroids.

Hagihara (1957), referring to the problem of the stability
of the solar system, arrived at the conclusion that the then
existing mathematical techniques and theoretical models could
not answer the question. This statement is still actual, but we
have made progress in these studies during the last 20 years.

We mentioned that the question of stability or instability
of planetary systems should be related to certain intervals of
time, and also the bodies that make up the system. In the case
of the solar system, we might consider it stable for a very
short period and referring to planets and certain satellites;
but the case might be different for asteroids and the particles
forming the rings around some planets, because they are far
more numerous and instabilities might set in more quickly.

When we think about short time intervals, the stability of
the planets' orbits might be possible to consider, but for long
and very long time intervals, the problem becomes complex, because our theoretical models do not take into account all the

possible variables acting for long periods of time.

We should really consider the physical nature of the bodies; also the existence of accelerations, due to non-gravitational forces and acting for long periods of time, might have an influence in the behaviour of the system. Tidal effects are worth considering, especially when we try to forecast the configuration of the system for longer periods of time.

The consideration of theoretical models including, at least, some of the physical variables mentioned, might give us a better insight into the problems of stability and instability of planetary systems, probably radically changing our views about the time intervals necessary for different degrees of instability to appear.

REFERENCES.

Hagihara, Y. (1957), "Stability in Celestial Mechanics",
     Kozai Press, Japan.

Jeffreys, H. (1967), "Theory of Probability", Oxford Univ. Press.

Laplace, P.S. (1830), "The System of the World", Translated by
     H.H. Harte, (Two Volumes), Univ. Press, Dublin, Ireland.

Siegel, C.L. (1956), "Vorlesungen uber Himmelsmechanik, Springer.

Szebehely, V. (1978), Cele. Mech. 17, In Press.

Szebehely, V. (1967), "Theory of Orbits", Academic Press.

Szebehely, V. and McKenzie, R. (1977), Astron. J. 82, pp. 303-
     305.

Thomson, W. and Tait, P.G. (1883), "Treatise on Natural Philoso-
     phy", Vol. 1, Part II, Cambridge Univ. Press.

PART IV

THE PROBLEM OF THREE BODIES

NOTE ON THE TRIANGULAR LAGRANGIAN POINTS IN THE ELLIPTIC RESTRICTED PROBLEM OF THREE BODIES

Hiroshi Kinoshita

Tokyo Astronomical Observatory, University of Tokyo, Japan

ABSTRACT. The general solutions of the infinitesimal motions around the equilateral triangular points in the elliptic restricted problem can be obtained by the perturbation method based on the averaging method provided the mass ratio of the primaries is smaller than the critical value $\mu_B = 1/2 - \sqrt{69}/18$. When $\mu \geq \mu_B$, the zeroth order solutions are not periodic. Ordinary perturbation methods, therefore, fail. In this note we discuss a new mthod which can treat the case where the zeroth order solutions are not periodic, and apply the method to the elliptic restricted problem.

1. INTRODUCTION.

It is well known that the restricted problem of three bodies has five particular solutions. Three of these are collinear and two are equilateral. In this note the triangular configuration is discussed. If the mass ratio and the eccentricity of the primaries satisfy certain conditions the triangular points are stable. Strictly speaking they are stable, but not secularily stable. In other words, if there is a non-conservative force like drag due to resisting medium, these points are not stable.

The motions around the triangular points are described by the following linear equations (Szebehely, 1967):

Victor G. Szebehely (ed.), Instabilities in Dynamical Systems. 229-241.
Copyright © 1979 by D. Reidel Publishing Company.

$$\ddot{q}_1 - 2\dot{q}_2 = gh_1 q_1 \, ,$$

$$\ddot{q}_2 + 2\dot{q}_1 = gh_2 q_2 \, ,$$

(1)

where $g = 1/(1+e \cos t)$ and $h_{1,2} = \frac{3}{2}[1 \pm \sqrt{1-3 \ (1-\mu)} \ .$

Here e is the eccentricity of the primaries, t is the true anomaly, and μ is the mass-ratio. The q_1 axis makes an angle $\alpha = \frac{1}{2} \arctan \sqrt{3}(1-2\mu)$ with respect to the line connecting the primaries. When the eccentricity is zero, these equations are reduced to linear differential equations with constant coefficients. The stability depends on the mass-ratio. If μ is betwen 0 and $\mu_B = \frac{1}{2} - \sqrt{69}/18 \doteq 0.03852$, the system is stable. The two frequencies ω_1, ω_2 of the stable systems are

$$\omega_{1,2}^{2} = \frac{1}{2} \ [1 \pm \sqrt{1-27\mu(1-\mu)} \, ,$$

$$0 < \omega_1 < 1/\sqrt{2} < \omega_2 < 1 \, ,$$

$$0 < \ \omega_1 + \omega_2 < 1 \ .$$

When $\mu = \mu_B$, the two frequencies are equal and the system is unstable.

Now we discuss the case $e \neq 0$. Equations (1) are linear differential equations with periodic coefficients. We cannot obtain the general solutions by quadrature. According to Floquet's theory, the general solutions of linear differential equations with periodic coefficients have the form,

$$\sum e^{\lambda_i t} X,$$

where λ_i are the characteristic exponents and $\rho_i = e^{\lambda_i}$ is the characteristic root. There is no method to obtain the characteristic exponents analytically. They may be obtained by a perturbation method if there is a small parameter. Characteristic

exponents or characteristic roots can be developed in power series
of this small parameter. In our problem the eccentricity is the
small parameter. Characteristic roots may be obtained numerically,
giving particular values for e and μ. First Danby (1964) obtained the characteristic roots, numerically, and later Bennett
(1965) caluculated them. In his Figure 1 at point A, one of the
frequencies is one half of the forced frequency and a 2:1 resonance occurs. The motion near this point was discussed by
Alfriend and Rand (1968), Nayfeh and Kamel (1970), Kinoshita
(1970), and Vinh (1972). At the point B, the two frequencies
are equal and an internal resonance takes place. The motion
near this point was discussed by Aoki (1955), and Vinh (1972).
The transition curve emanating from A is given by

$$\mu = \frac{1}{2} - \frac{\sqrt{2}}{3} \pm \frac{\sqrt{66}}{144} \, e + \frac{49\sqrt{2}}{4608} \, e^2 \pm \frac{751\sqrt{66}}{270336} \, e^3 - \frac{114275\sqrt{2}}{14155776} \, e^4 + 0(e^5).$$

$$(2)$$

The transition curve emanating from B is given by

$$\mu = \frac{1}{2} - \frac{\sqrt{69}}{18} \, [1 - \frac{4}{23} \, e^2 + \frac{2305}{4232} \, e^4] + 0(e^6). \tag{3}$$

These analytical expressions were obtained by Vinh (1972). The
interesting feature is that the transition curve emanating from
B is curved towards the right. The system with e=0 and
$\mu > \mu_B$ is unstable. However, even if the mass ratio is larger

than μ_B, the system becomes stable provided the eccentricity

is large enough.

Analytical expressions of the characteristic exponents up
to e^2 were obtained by Deprit and Rom (1970) and Vinh (1972).
Vinh (1972) did not take into account the case $\mu = \mu_B$.

Canonical transformations.

The perturbation method based on canonical transformation
is discussed which can be applied to the system whose zeroth
order solutions are not periodic.

The Hamiltonian for the differential equation (1) is

$$H = \frac{1}{2} \, (p_1 + q_2)^2 + \frac{1}{2} \, (p_2 - q_1)^2 - \frac{1}{2} \, g \, (h_1 q_1^{\,2} + h_2 q_2^{\,2}) \tag{4}$$

where $p_1 = \dot{q}_1 - q_2$ and $p_2 = \dot{q}_2 + q_1$, and p_1 and p_2 are conjugate to q_1 and q_2. This can be written in a quadratic form as

$$H = \frac{1}{2} x'Kx , \qquad (5)$$

where

$$K = \begin{pmatrix} i-ga & -j \\ j & i \end{pmatrix}$$

with

$$i = \begin{pmatrix} 1 & 0 \\ 0 & 1 \end{pmatrix}, \quad j = \begin{pmatrix} 0 & 1 \\ -1 & 0 \end{pmatrix} \quad a = \begin{pmatrix} h_1 & 0 \\ 0 & h_2 \end{pmatrix}$$

and $x' = (q_1 q_2 p_1 p_2)$. The prime attached to vectors or matrices represents the transpose. K is a 4x4 matrix.

Let us consider a canonical transformation defined by the following generating function S;

$$S = \frac{1}{2} x*'Wx* \qquad (6)$$

where W is an arbitrary symmetric matrix, which is generally a function of t. The relationship between x and x* is

$$x = e^{JW}x* \qquad (7)$$

where

$$J = \begin{pmatrix} 0 & i \\ -i & 0 \end{pmatrix} .$$

The derivation of (7) is given in Appendix 1. The canonical transformation (7) is linear and symplectic. With this transformation the new Hamiltonian takes the following form:

$$H* = \frac{1}{2} x*'K*x* , \qquad (8)$$

where

$$K^* = e^{-JW} + [J \frac{d}{dt} (e^{JW})]'e^{JW}. \qquad (9)$$

Note the appearance of $\frac{d}{dt}$ in K^*, due to the fact that the transformation S depends on time. The derivation of Equation (9) is given in Appendix 2. The differential equation for the new coordinates x^* is

$$\frac{dx^*}{dt} = JK^*x^* . \qquad (10)$$

If K^* is a constant matrix, the fundamental matrix is written as

$$X = e^{JK^*t} . \qquad (11)$$

The general solution of (10) is

$$x^* = XC = e^{JK^*t}C$$

where C is an arbitrary constant vector, which depends on the initial conditions.

We have only one equation for W and K^*. Therefore we have one degree of freedom to determine W and K^*. We choose W by requiring that K^* be a constant. Expanding K^* and W in power series of the small parameter and equating the same order terms with respect to the small parameter, we have

$$K^*_o = K_o ,$$

$$K^*_1 = - \frac{dW_1}{dt} + K_o JW_1 - W_1 JK_o + K_1 ,$$

$$K^*_2 = - \frac{dW_2}{dt} + K_o JW_2 - W_2 JK_o + K_2 + R_2 , \qquad (13)$$

$$\cdot$$
$$\cdot$$
$$\cdot$$

where $R_2 = \frac{1}{2} [(K_1 + K^*_1)JW_1 - JW_1(K_1 + K^*_1)]$.

Using the condition that K^* must be constant, we have from (13)

$$K_1^* = K_{1 \text{ sec}} \text{ ,}$$

$$W_1 = \int^t e^{-K_o J(s-t)} K_{1p} e^{JK_o(s-t)} ds \text{ ,}$$

$$K_2^* = [K_2 + R_2] \text{ sec ,}$$

$$W_2 = \int^t e^{-K_o J(s-t)} [K_{2p} + R_{2p}] e^{JK_o(s-t)} ds \text{ ,}$$

.
.
.

(14)

Note that W_n is a periodic function with respect to the period of the forced term, and not with respect to the frequencies of the zeroth order solution.

Let us consider a constant matrix, A, whose eigenvalues are $\pm i\,\omega_1$ and $\pm i\,\omega_2$. First we consider the case $\omega_1 \neq \omega_2$.

Using the Lagrange-Sylvester interpolation polynomial (see Gantmacher, 1959), we have

$$e^{At} = A_1 e^{i\omega_1 t} + A_2 e^{-i\omega_1 t} + A_3 e^{i\omega_2 t} + A_4 e^{-i\omega_2 t} \tag{15}$$

where

$$A_{1,2} = \frac{1}{2(\omega_1{}^2 - \omega_2{}^2)} \left(-E_2 \pm \frac{iAE_2}{\omega 1} \right),$$

$$A_{3,4} = \frac{1}{2(\omega_2{}^2 - \omega_1{}^2)} \left(-E_1 \pm \frac{iAE_1}{\omega_2} \right), \tag{16}$$

with

$$E_1 = A^2 + \omega_1{}^2 I \text{ ,}$$

$$E_2 = A^2 + \omega_1{}^2 I \text{ .}$$

We calculate the following integral which appears in (14):

$$\int^t e^{A'(s-t)} D\, e^{A(s-t)} \begin{Bmatrix} \cos \\ \sin \end{Bmatrix} ms\, ds \ ,$$

where D is a constant matrix and m is a positive integer. The following was obtained by Vinh (1972):

$$\int^t e^{A'(s-t)} D\, e^{A(s-t)} \begin{Bmatrix} \cos \\ \sin \end{Bmatrix} ms\, ds \ ,$$

$$= \begin{Bmatrix} \sin\ mt \\ -\cos\ mt \end{Bmatrix} P_m(D) + \begin{Bmatrix} \cos\ mt \\ \sin\ mt \end{Bmatrix} Q_m(D) \ ,$$

where

$$m(m^2 - 4\omega_1^2)(m^2 - 4\omega_2^2)[(m^2-1)^2 - 4\omega_1^2\omega_2^2]P_m(D) =$$

$$= [m^2(m^2-1)^2(m^2-4) + 2m^2(5m^2-3)\omega_1^2\omega_2^2 + 8\omega_1^2\omega_2^2(1-\omega_1^2\omega_2^2)]D$$

$$-2[(m^2-1)^2(m^2-4) + 6(m^2+2)\omega_1^2\omega_2^2]A'DA$$

$$+2[3m^2(m^2-1) + 8\omega_1^2\omega_2^2)] (A')^2DA^2$$

$$-4(5m^2-2)(A')^3 DA^3 \hspace{4cm} (18\text{-}1)$$

$$-[m^2(m^2-1)(m^2-4) - 4(m^2+2)\omega_1^2\omega_2^2]B_2$$

$$+2[(2m^2-1)(m^2-4) - 8\omega_1^2\omega_2^2]A'B_2A, \quad \text{and}$$

$$(m^2-4m^2)(m^2-4\omega_2^2)[(m^2-1)^2 - 4\omega_1^2\omega_2^2)] Q_m(D)$$

$$= 2(5m^2-2)(A')^2B_1A^2$$

$$+[(m^2-1)^2(m^2-4) + 6(m^2+2)\omega_1^2\omega_2^2]B_1$$

$$-[(m^2-1)(m^2-4) - 24\omega_1^2\omega_2^2]B_3$$

$$-[(3m^2-1)(m^2-4) + 8\omega_1\omega_2^2]A'B_1A \ ,$$

with

$$B_1 = A'D + DA ,$$

$$B_2 = (A')^2D + DA^2 ,$$

$$B_3 = (A')^3D + DA^3 .$$

Next we consider the case $\omega_1 = \omega_2 = \omega$, which Vinh did not investigate. We have

$$e^{At} = A_1 e^{i\omega t} + A_2 e^{-i\omega t} + A_3 t e^{i\omega t} + A_4 t e^{-i\omega t} \qquad (26)$$

$$A_{1,2} = \frac{1}{2} \pm \frac{i(3A\omega^2+A^3)}{4\omega_3} ,$$

$$A_{3,4} = \pm \frac{i(A^2+\omega^2)}{4\omega} - \frac{A(A^2+\omega^2)}{4\omega^2} . \qquad (21)$$

Now we return to the original problem. Putting

$$A=JK \begin{pmatrix} i & i \\ a-i & j \end{pmatrix} , \quad D= \begin{pmatrix} a & o \\ o & u \end{pmatrix} , \quad m=1 \text{ and using (17), we have}$$

from (14)

$$W_1 = e(P_1 \sin t + Q_1 \omega t) \qquad (22)$$

where

$$P_1 = \frac{1}{16d-3} \begin{pmatrix} -11a+8da+10di+24i , & -ja-11aj+8dj \\ aj+11ja-8dj , & 14a-8di-36i \end{pmatrix} , \quad (23)$$

$$Q_1 = \frac{1}{16d-3} \begin{pmatrix} 0 & 6i+a+4di \\ 6i+a+4di , & 4(aj-ja) \end{pmatrix} , \qquad (24)$$

with $d = \omega_1^2 \omega_2^2$,

and

$$K_1^* = 0 , \tag{25}$$

$$K_2^* = \frac{e^2}{2(16d-3)} \begin{pmatrix} 12a-12ad-di & , & 2(3a-2di)j \\ \\ -2j(3a-2di) & , & 0 \end{pmatrix} . \tag{26}$$

The secular equation or the characteristic equation for (10) is

$$\Delta = |\lambda I - JK^*|$$

$$= \lambda^4 + (1 - \frac{11d}{16d-3} e^2)^2 + d \left[1 + \frac{3(8d-1)}{(16d-3)} e^2 \right] +0(e^4) , \tag{27}$$

where $d = \omega_1^2 \omega_2^2$.

In order for the system to be stable, the discriminant of (27) must be positive:

$$G = 1-4d + \frac{d(20-48d)}{16d-3} e^2 \geq 0 . \tag{28}$$

This relation determines the transition curve emanating from B and it agrees with that obtained by Aoki (1955) and Vinh (1972).

The general solution of the original equation (1) has the following form:

$$x = e^{JW(t)} e^{JK^* t} C . \tag{29}$$

This form agrees with that predicted from the Floquet's theorem. The characteristic exponents are the roots of $\Delta=0$:

$$\lambda = i\omega^* ,$$

$$\omega_{1,2}^{*2} = \frac{1}{2} [1 - \frac{11d}{16d-3} e^2 \pm \sqrt{G}] + 0(e^4) \text{ for } G>0 \tag{30}$$

and, when $\mu < \mu_B$ and $\mu \neq \mu_A$,

$$\omega_i^* = \omega_i [1 + \frac{(1-\omega_i^2)(7-6\omega_i^2)}{4(1-2\omega_i^2)(1-4\omega_i^2)} e^2] + 0(e^4) \text{ for } i=1.2 . \tag{31}$$

The equation (31) coincides with Deprit's (1970) result, but
Deprit's procedure cannot produce (30) because of the reason
mentioned before.

Appendix 1

Let q_i^* and p_i^* be a set of $2n$ canonical variables
and $f(q*,p*)$ and $S(q*,p*)$ be arbitrary functions of $q*$
and $p*$. Operators D_S^n are defined by

$$D_S^\circ f = f \; , \quad D_S' f = \{f,S\} \; , \quad D_S^n f = D_S^{n-1}\{D_S^1 f\} \; , \quad n \geq 2 ,$$

where braces stand for Poisson's brackets. The theorem of Lie
(1888) states that a set of $2n$ variables q_i, p_i defined by
the equation

$$f(p,q) = \sum_{n=0}^{\infty} \frac{1}{n!} D_S^n f(q*,q*) \tag{A-1}$$

is canonical if the series in the right-hand side of (A-1) con-
verges (see Hori's paper 1966).

Now let us consider the following case

$$x* = \begin{pmatrix} q* \\ p* \end{pmatrix} ,$$

$$f(x*) = Ax* ,$$

$$S = \frac{1}{2} x*'W x* ,$$

where A is an arbitrary matrix which does not depend on $q*$
and $p*$, W is a symmetric matrix, and S is a quadratic form.
The Poisson barcket of S and of the i-th component of f is

$$\{f_i, S\} = \left(\sum_{\alpha=1}^{n} \frac{\partial f_i}{\partial x^*} \frac{\partial S}{\partial x^*_{n+\alpha}} - \frac{\partial f_i}{\partial x^*_{n+\alpha}} \frac{\partial S}{\partial x^*} \right)$$

$$= \sum_{\alpha=1}^{n} [A_{i,\alpha} (Wx^*)_{n+\alpha} - A_{i,n+\alpha} (Wx^*)_{\alpha}]$$

$$= \sum_{\alpha=1}^{n} \sum_{j=1}^{2n} (A_{i,\alpha} W_{n+\alpha,j} x^*_j - A_{i,n+\alpha} W_{\alpha,j} x^*_j)$$

$$= \sum_{j=1}^{2n} [\sum_{\alpha=1}^{n} (A_{i,\alpha} W_{n+\alpha,j} - A_{i,n+\alpha} W_{\alpha,j})] x^*_j$$

$$= \sum_{j=1}^{2n} \sum_{k=1}^{2n} A_{i,k} (JW)_{k,j} x^*_j = \sum_{j=1}^{2n} (AJW)_{ij} x^*_j$$

$$= (AJWx^*)i .$$

Therefore we have

$$\{Ax^*, S\} = AJWx^* .$$ (A-2)

Using (A-2) we obtain

$$\{AJWx^*, S\} = AJWJWx^* = A(JW)^2 x^* .$$

Therefore

$$Ax = \sum_{n=0}^{\infty} \frac{1}{n_i} D_S^n Ax^*$$

$$= Ax^* + AJWx^* + \frac{1}{2_i} A(JW)^2 x^* + \ldots$$

$$= A(I + JW + \frac{1}{2_i} (JW)^2 + \ldots) x^*$$

$$= Ae^{JW} x^* .$$

Now we have an explicit relationship between x and x*;

$$x = e^{JW} x^*$$

which is a canonical transformation and symplectic. We can prove that e^{JW} is symplectic by showing directly that

$$e^{-WJ} e^{JW} = J \quad \text{or} \quad Je^{JW} = e^{WJ}J \quad \text{using the Taylor expansion}$$

theorem.

Appendix 2

Let us consider a linear canonical transformation defined by

$$\begin{pmatrix} q \\ p \end{pmatrix} = T \begin{pmatrix} Q \\ P \end{pmatrix} = \begin{pmatrix} A & B \\ C & D \end{pmatrix} \begin{pmatrix} Q \\ P \end{pmatrix} , \qquad (A-4)$$

where q and p are n-dimentional vectors, A,B,C, and D are nxn matrices, and p is the canonically conjugate of q. The matrix T generally depends on the time t. The symplectic matrix T satisfies the equations,

$$TJT' = J \quad \text{or} \quad T'JT = J . \qquad (A-5)$$

From (A-5) we have explicit relations among A,B,C, and D:

$$AB' = BA' \quad , \quad D'B = B'D \quad ,$$

$$CD' = DC' \quad , \quad A'C = C'A \quad , \qquad (A-6)$$

$$AD'-BC'=I \quad , \quad A'D-C'B=I \quad ,$$

where I is a unit matrix. By using (A-4), vectors q and Q are expressed in terms of p and P as

$$q = AC^{-1}p - (C')^{-1}P \quad ,$$

$$Q = C^{-1}p - C^{-1}DP \quad . \qquad (A-7)$$

Let us determine the generating function $F(p,P)$ for the canonical transformation (A-4):

$$q = - \frac{\partial F}{\partial p} \quad ,$$

$$Q = \frac{\partial F}{\partial P} \quad . \qquad (A-8)$$

Noticing that AC^{-1} and $C^{-1}D$ are symmetric matrices and using (A-7) and (A-8), we can express F in a quadratic form:

$$F = -\frac{1}{2}P' \ AC^{-1}p - \frac{1}{2}P'C^{-1}DP + P'C^{-1}p \ . \qquad (A-9)$$

Taking the partial derivative of F, and with respect to time and substituting p=AQ+CP,

$$\frac{F}{t} = \frac{1}{2}(Q'P')[J \ \frac{d}{dt} \ \begin{pmatrix} A & B \\ C & D \end{pmatrix}]' \ \begin{pmatrix} A & B \\ C & D \end{pmatrix} \ \begin{pmatrix} Q \\ P \end{pmatrix} \qquad (A-10)$$

If we put $x = \begin{pmatrix} q \\ p \end{pmatrix}$, $x^* = \begin{pmatrix} Q \\ P \end{pmatrix}$, and $T=e^{JW}$, we obtain Equation (9).

2. ACKNOWLEDGEMENTS.

This study was prompted by Vinh's paper (1972) whose notation was accepted. The author would like to express his thanks to Drs. Hori and Henrard for their valuable discussions.

3. REFERENCES.

Alfriend, K.T. and Rand, R.H. (1968), J. Astronaut. Sci. 15, p. 105.

Aoki, S. (1955) Publ. Astr. Soc. Japan 7, p. 105.

Bennett, A. (1965), Icarus 4, p. 177.

Danby, J.M.A. (1964), Astr. J. 69, p. 165.

Deprit, A. and Rom, A. (1970), Astron. & Astrophys. 5, p. 416.

Gandmacher, F.R. (1959), Matrix Theory, Chelsea Pub. Comp., vol. 1, pp. 101-103.

Hori, G. (1966), Publ. Astr. Soc. Japan 18, p. 287.

Kinoshita, H. (1970), Publ. Astr. Soc. Japan 22, p. 373.

Nayfeh, A.H. and Kamel, A.A. (1970), Am. Inst. Aeronaut. Astronaut. J. 8, p. 221.

Szebehely, V. (1967), "Theory of Orbits", Academic Press, pp. 254-255.

Vinh, N.X. (1972), Cel. Mech. 6, p. 305.

PERIODIC ORBITS WITH AND WITHOUT SYMMETRY IN THE THREE-DIMENSIONAL GENERAL THREE-BODY PROBLEM

Vassilis Markellos

University of Glasgow, Glasgow, Scotland

ABSTRACT. We report on the first results of a numerical exploration of the periodic orbits of the three-dimensional general
three-body problem. The problem is formulated in a moving coordinate system appropriate for the continuation of the known
periodic orbits of the three-dimensional restricted problem to
the general problem. Periodic orbits of the three-dimensional
general problem are determined and a procedure for their global
exploration is outlined.

This is a short report on a large scale numerical exploration of the periodic solutions of the three-dimensional general three-body problem undertaken in the last year at the University of Glasgow.

The three-dimensional problem is formulated in a moving
coordinate system suitable for the continuation of periodic orbits from the three-dimensional restricted problem. The origin
of the system is taken to be the center of mass of the 'primaries'
m_1 and m_2 and the positive x-axis is directed from m_1

towards m_2. In this system the positions of the two primaries

are fully determined if the distance from the origin of one of
them is known. The state of motion of the full three-body system is described by six variables: $x, y, z, x_2, \theta, \phi$, and their

time derivatives. The coordinates x, y, z describe the position
of the third body m_3 in the moving system and the angles

Victor G. Szebehely (ed.), Instabilities in Dynamical Systems. 243-250.
Copyright © 1979 by D. Reidel Publishing Company.

θ and ϕ specify the orientation of the x-axis in the three-dimensional space, relative to a fixed system of reference. The four coordinates: x,y,z,x_2 fully describe the position of all

three bodies in the moving system, and if the two angles are also known then the positions of the three bodies in the fixed system can also be determined. The formulation is given by the equations of motion in Table I. In this formulation we take the gravitational constant $k^2 = 1$, we normalize the total mass of the system: $m_1 + m_2 + m_3 = 1$, and use the notation

$\mu = m_2/(m_1 + m_2)$.

The periodicity conditions for (relative) periodic motion to occur in the moving coordinate system are easily formulated for general periodic motion (without symmetry) and can be reduced in number by the existence of the integrals of energy and angular momentum. The relevant expressions for the integrals are given explicitly in Table II, and the periodicity conditions are summarized in Table III. The equations of motion of Table I have certain symmetry properties which result in the possibility of establishing three-dimensional periodic orbits of the general three-body problem possessing symmetries of the same types as do the periodic orbits of the three-dimensional restricted three-body problem. The simplified periodicity conditions for symmetric periodic orbits of the three-dimensional general problem are also summarized in Table III.

Based on the periodicity conditions of Table III suitable differential corrections and iterative procedures can be established which give the initial conditions of periodic orbit, with or without symmetries, once the appropriate starting values for the iterative procedure are given. An appropriate normalization scheme is introduced based on the concept of the "invariable" plane. In this scheme the x and y components of the angular momentum vector are always zero. A fuller report is in the process of publication elsewhere.

To proceed with a large scale numerical determination of all the (basic) families of periodic orbits of the three-dimensional general problem we have two options. Either we can systematically extend the known families of the three-dimensional restricted problem (and the present formulation of the general problem is designed with such numerical continuation in mind), or we can extend the known families of the plane general problem. Neither option is very promising due to lack of extensive results at the present time concerning either the three-dimensional restricted problem or the plane general problem.

TABLE I

THREE-DIMENSIONAL GENERAL THREE-BODY PROBLEM

EQUATIONS OF MOTION

IN THE MOVING COORDINATE SYSTEM.

$$\ddot{x} = (B+\dot{\phi}^2) x + \cos\phi \ (\alpha\dot{\theta}^2+2\dot{\theta}\dot{y}+\ddot{\theta}y) + 2\dot{\phi}\dot{z} + \ddot{\phi}z + \mu A x_2$$

$$\ddot{y} = (B+\dot{\theta}^2) y - \alpha\ddot{\theta} - 2\beta\dot{\theta} + 2\lambda\dot{\theta}\dot{\phi}$$

$$\ddot{z} = (B+\dot{\phi}^2) z - \sin\phi \ (\alpha\dot{\theta}^2+2\dot{\theta}\dot{y} +\ddot{\theta}y) - 2\dot{\phi}\dot{x} -\ddot{\phi}x$$

$$\ddot{x}_2 = (m_3 B*+\dot{\theta}^2\cos^2\phi+\dot{\phi}^2)x_2 - \frac{(1-m_3)(1-\mu)^3}{x_2^2} + m_3(1-\mu)Ax$$

$$\ddot{\theta} \ x_2\cos\phi = 2\dot{\theta}\dot{\phi}x_2\sin\phi - 2\dot{\theta}\dot{x}_2\cos\phi + m_3(1-\mu)Ay$$

$$\ddot{\phi} \ x_2 = - \dot{\theta}^2 x_2\cos\phi \ \sin\phi - 2\dot{\phi}\dot{x}_2 + m_3(1-\mu)Az$$

$$B = - \left(\frac{1-\mu}{r_{13}^3} + \frac{\mu}{r_{23}^3} \right), \quad B* = - \left(\frac{\mu}{r_{13}^3} + \frac{1-\mu}{r_{23}^3} \right), \quad A = - \left(\frac{1}{r_{13}^3} - \frac{1}{r_{23}^3} \right)$$

$$r_{13} = [(x+\frac{\mu}{1-\mu} x_2)^2+ y^2 + z^2]^{1/2}, \quad r_{23} = [(x-x_2)^2+ y^2 + z^2]^{1/2}.$$

$$\alpha = x \cos\phi - z \sin\phi, \quad \beta = \dot{x} \cos\phi - \dot{z} \sin\phi, \quad \lambda = x \sin\phi + z \cos\phi.$$

TABLE II

ENERGY INTEGRAL $\qquad$ $E = T + V$

$$T = 1/2 \ (1-m_3) \ (\frac{\mu}{1-\mu} \ \dot{r}_2^2 + m_3 \ \dot{r}_3^2)$$

$$V = - \frac{\mu(1-\mu)^2(1-m_3)^2}{x_2} - m_3(1-m_3) \ (\frac{1-\mu}{r_{13}} + \frac{\mu}{r_{23}})$$

$$\dot{r}_3^2 = \dot{x}^2+\dot{y}^2+\dot{z}^2+(\dot{\alpha}^2+y^2)\dot{\theta}^2+(x^2+z^2)\dot{\phi}^2+2\dot{\theta}(\alpha\dot{y}-\beta y)+2\dot{\phi}(x\dot{a}-\dot{x}z)+2\lambda\dot{\theta}\dot{\phi}y$$

$$\dot{r}_2^2 = \dot{x}_2^2 + x_2^2 \ (\dot{\theta}^2\cos^2\phi+\dot{\phi}^2)$$

ANGULAR MOMENTUM INTEGRAL (z component)

$$P_\theta = A_3 = \frac{(1-m_3)}{1-\mu} \ x_2^2 \ \dot{\theta} \ \cos^2\phi + m_3(1-m_3) \ [(\alpha^2+y^2)\dot{\theta}+\alpha\dot{y}-\beta y+\lambda\dot{\phi}y]$$

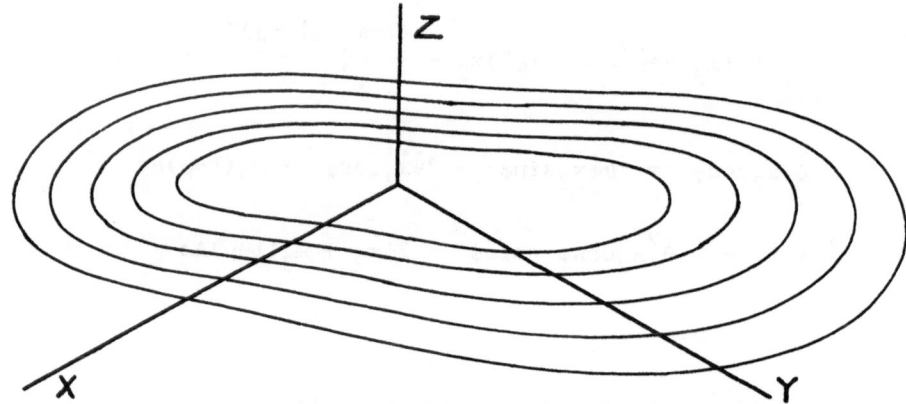

Figure 1. A family of periodic orbits of the three-dimensional general three-body problem determined by numerical continuation of a periodic orbit of the circular restricted problem. The starting orbit is a member of the family C2v of the restricted problem which bifurcates into three-dimensions from the family c of plane orbits around the inner Lagrangian equilibrium point. That bifurcation occurs at the vertical-critical orbit c2v of the restricted problem (Hénon, 1974). The orbits of the present family all have the same starting value of the z-component of the velocity vector $\dot{z}_0 = 0.5$. The two primaries are assumed equal ($\mu = 0.5$) but the third mass m_3 varies along this family. The orbits increase in size as m_3 increases and the ones illustrated here correspond to the values 0.1, 0.3, 0.5, 0.7 and 0.9 of m_3. The family is called C2v and consists of axisymmetric orbits.

<u>TABLE III</u>

<u>PERIODICITY CONDITIONS</u>

<u>General case</u> : <u>Non-symmetric orbits</u>

$$x \quad (x_o, z_o, x_{20}, \dot{x}_o, \dot{y}_o, \dot{z}_o, \dot{x}_{20}, T, m_3) = x_o$$

$$z \quad (\qquad\qquad\qquad\qquad\qquad) = z_o$$

$$x_2 \quad (\qquad\qquad\qquad\qquad\qquad) = x_{20}$$

$$\dot{x} \quad (\qquad\qquad\qquad\qquad\qquad) = \dot{x}_o$$

$$\dot{y} \quad (\qquad\qquad\qquad\qquad\qquad) = \dot{y}_o$$

$$\dot{z} \quad (\qquad\qquad\qquad\qquad\qquad) = \dot{z}_o$$

$$\dot{x}_2 \quad (\qquad\qquad\qquad\qquad\qquad) = \dot{x}_{20}$$

<u>Special cases</u> : <u>Symmetric orbits</u>

Plane symmetric orbits

$$y \quad (x_o, z_o, x_{20}, \dot{y}_o; T/2, \mu, m_3) = y_o = 0$$

$$\dot{x} \quad (\qquad\qquad\qquad\qquad) = \dot{x}_o = 0$$

$$\dot{z} \quad (\qquad\qquad\qquad\qquad) = \dot{z}_o = 0$$

$$\dot{x}_2 \quad (\qquad\qquad\qquad\qquad) = \dot{x}_{20} = 0$$

Axisymmetric orbits

$$y \quad (x_o, x_{20}, \dot{y}_o, \dot{z}_o, T/2, \mu, m_3) = y_o = 0$$

$$z \quad (\qquad\qquad\qquad\qquad) = z_o = 0$$

$$\dot{x} \quad (\qquad\qquad\qquad\qquad) = \dot{x}_o = 0$$

$$\dot{x}_2 \quad (\qquad\qquad\qquad\qquad) = \dot{x}_{20} = 0$$

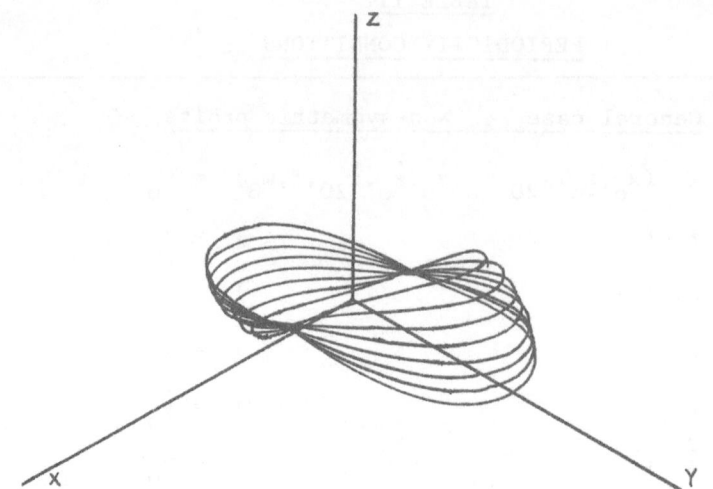

Figure 2. The family C2v for fixed (equal) masses ($\mu = 0.5$, $m_3 = 1/3$) and varying inclination.

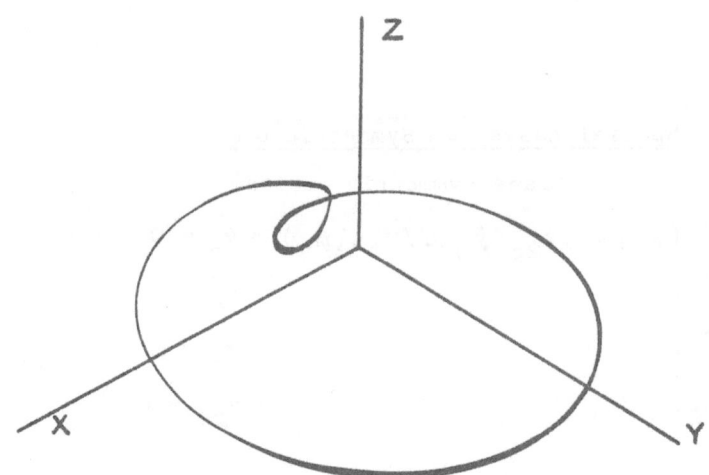

Figure 3. A family of nonsymmetric periodic orbits of the three-dimensional general problem determined by numerical continuation of a periodic orbit of the restricted problem. The starting orbit is a member of the family Al given in Markellos (1978). The orbits illustrated correspond to $\mu = 0.5$, for m_3 varying in the interval (0, 1/20). In this and the previous figures the orbit of the third body m_3 is shown. The orbits of the primaries m_1 and m_2 in the moving coordinates employed are simply oscillations on the x-axis and would appear in these figures as intervals of that axis.

Thus a preliminary phase of numerical exploration is required to produce on a large scale the starting points of the basic families of periodic orbits of the three-dimensional general problem. These are the initial conditions of the planar periodic orbits of the general problem which are "critical" with respect to their vertical stability. They represent bifurcations of families of plane periodic orbits to families of three-dimensional periodic orbits. Indeed, these critical orbits are available in some cases of the restricted problem (Henon, 1974; Markellos, 1977), and they may be continued numerically into the general problem in a systematic fashion. This approach has been adopted and many series of critical orbits of the general problem have been obtained. The method and the results are in the process of publication elsewhere. The approach has the distinct advantage that it allows predictions to be made about the stability of the three-dimensional periodic orbits of the general problem that bifurcate from the critical orbits.

While proceeding with large scale computation of the critical orbits of the general problem we have also obtained many families of three-dimensional periodic orbits of the general problem in the complete range of the mass parameters: $0 \le m_3 \le 1$ and $0 \le \mu \le 1/2$. Data for such families as well as the computer codes used are available upon request from the author. Here we give computer plots of three-dimensional orbits for illustration.

We note that the periodic orbits given or referred to here are periodic in general, only relative to the moving coordinate system. However, the monitoring of a single parameter can reveal the existence of many orbits which are also periodic relative to a fixed coordinate system. This parameter is $\theta(T)$ where T is the period. As $\theta(T)$ varies along a family of three-dimensional periodic orbits of the general problem, infinitely many values are met which are commensurable with π. Thus, if we impose no limit on the allowed periods in the fixed system then along any family of relative periodic orbits we can detect infinitely many orbits which are also periodic in the fixed system.

The families of orbits shown in Figures 1, 2 and 3 may be discussed as follows:

Figure 1 shows a family of periodic orbits of the three-dimensional general three-body problem determined by numerical continuation of a periodic orbit of the circular restricted problem. The starting orbit is a member of the family C2v of the restricted problem which bifurcates into three-dimensions from the family c of plane orbits around the inner Lagrangian point. This bifurcation occurs at the vertical-critical orbit c2v of the restricted problem (Henon, 1974). The orbits of the family shown have the same starting value regarding the z-component of the velocity vector, $\dot{z}_o$ = 0.5. The two primaries have

equal mass (μ = 0.5) and the third mass, m_3 varies along the

family. The orbits increase in size as m_3 increases and the

ones illustrated here correspoind to the values 0.1, 0.3, 0.5, 0.7 and 0.9 of m_3. The family is called C2v and consists of axisymmetric orbits.

Figure 2 shows the same family with fixed (equal) masses (μ = 0.5, m_3 = 1/3) and varying inclination. This family was

determined by numerical continuation of the vertical-critical orbit, c2v of the restricted problem to the general problem. Then the bifurcation c2v plane orbit was used, corresponding to the above values of the masses as the starting point.

Figure 3 shows a family of nonsymmetric periodic orbits of the three-dimensional general problem determined by numerical continuation of a periodic orbit of the restricted problem. The starting orbit is a member of the family A1 given in Markellos (1978). The orbits illustrated correspond to μ = 0.5 and to $0 \le m_3 \le 1/20$. In this and in the previous figures, the orbit

of the third body, m_3 is shown. The orbits of the "primaries"

m_1 and m_2 in the moving coordinates are oscillations on the

x-axis.

REFERENCES.

Henon, M. (1974), Astron. Astrophys. 28, p. 415.

Markellos, V.V. (1977), Mon. Not. R. Astr. Soc. 180, p. 103.

Markellos, V.V. (1978), Mon.Not.R. Astr. Soc. 184, p. 273.

THE VICINITY OF THE PRIMARY OF SMALL MASS
IN THE RESTRICTED PROBLEM

F. Nahon
University of Paris

ABSTRACT. If r is distance of the moving mass M to the primary
of small mass M_2 and is sufficiently small, then the orbits are
Keplerian orbits with respect to M_2, perturbed by the large but
distant mass M_1.

   We give an analytical study valid for all time of two families of orbits: (1) the collision orbits, and (2) the perturbed
orbits, using in the first case an adaptation of Levi-Civita's
(1903) method, and in the second a simplification of a more
general study by Kevorkian (1962) for a satellite problem.

1. INTRODUCTION

   The Hamiltonian of the restricted problem belongs to the class

$$F = F_o + \mu \, F_1,$$

where F_o is the Hamiltonian of an integrable problem (Keplerian
motion in a rotating system), μ is the small parameter = the
mass of the second primary, and F_1 is the perturbation. This
term contains the singularity $r_2 = 0$ which is not a singularity
for F_o. Therefore, if r_2 is small enough, the restricted problem
is a Keplerian problem with respect to the small mass M_2, perturbed by the large but distant mass M_1. We do not define the
"radius of action" $r_2(\mu)$, but study two families of orbits: the
collision orbits and the perturbed circular orbits.

Victor G. Szebehely (ed.), Instabilities in Dynamical Systems. 251-262.

Part 1.   The collision orbits

Levi-Civita's method, the proposed method and the results are described.   In this section consider the Hamiltonian in heliocentric coordinates

$$F(M_1) = \frac{1}{2} (R^2 + \frac{\Theta^2}{r^2}) - \Theta - \Omega(r,\theta)$$

$$\Omega = \frac{\nu}{r} + \frac{\mu}{\Delta} - \mu r \cos \theta \tag{1}$$

where Δ is the distance to M_2, and $\nu = 1 - \mu$.

If $\mu = 0$, the collision orbits are the Keplerian collision orbits in the rotating system and their representation is

$$\theta = \theta_o - t \qquad \Theta = 0$$

$$r = \rho^2 \qquad\qquad t = - \rho^3 a^{-3/2} [1 + 0(e)] \tag{2}$$

$$a^3 = \frac{9}{2}$$

In the general case, Levi-Civita eliminates R, using the Jacobian integral

$$R = \bar{R}[r,\theta,\Theta,C]$$

$$\frac{\bar{R}^2}{2} = \Omega - C + \Theta - \frac{1}{2} \frac{\Theta^2}{r^2} \tag{3}$$

Note that C is not the Szebehely's Jacobian constant since C refers to heliocentric coordinates.   Let $C(0)$ be the barycentric Jacobian constant.   Then

$$C = C(0) - \frac{1}{2} \text{ and } C(0) = \frac{1}{2} C_s,$$

where C_s is the Jacobian constant used by Szebehely (1967).

Returning to Levi-Civita, he proves that the collision orbits verifies an invariant relation

$$\Theta = \rho^5 f(\rho,\theta), \tag{4}$$

where f is the conic solution expandable in powers of $\rho = \sqrt{r}$ of the partial differential equation

$$5f + \rho \frac{\partial f}{\partial \rho} = \Phi(\rho,\theta,f, \frac{\partial f}{\partial \theta},\bar{R}). \tag{5}$$

The function Φ of course is known but very intricate because it depends on f through $\bar{R}$. Nevertheless, Levi-Civita gives the first terms of the expansion up to ρ^3.

Note that this study is perhaps the first example of regularization. Indeed, the use of $\rho = \sqrt{r}$ is essentially a regularization.

The proposed method. Consider the collision orbits with the primary M_2. The Hamiltonian in Jovicentric coordinates is

$$F(M_2) = \frac{1}{2} (R^2 + \frac{\Theta^2}{r^2}) - \Theta - \Omega$$

$$\Omega = \frac{\mu}{r} + \frac{\nu}{\Delta} + \nu r \cos \theta \tag{6}$$

where Δ is now the distance to M_1.

Proposition 1: To the family of collision orbits for a given value of the Jacobian constant, there corresponds a solution of the Hamilton-Jacobi partial differential equation:

$$\frac{1}{2} [(\frac{\partial s}{\partial r})^2 + \frac{1}{r^2} (\frac{\partial s}{\partial \theta})^2] - \frac{\partial s}{\partial \theta} = \Omega - C. \tag{7}$$

This solution is the conic solution expanded in powers of ρ in the vicinity of $\rho = 0$.

If $\mu \to 0$, the collision orbits exist only for $C(0) \leq \frac{3}{2}$ and for $C(0) < \frac{3}{2}$ they can be approximated by Keplerian orbits which join at M_2. The critical case is $C(0) = \frac{3}{2}$. Following Hill, we study this case by writing $r = \mu^{1/3} \bar{r}$ and considering Hill's Hamiltonian:

$$H = \frac{1}{2} (\bar{R}^2 + \frac{\bar{\Theta}^2}{\bar{r}^2}) - \bar{\Theta} - \bar{\Omega} ,$$

$$\bar{\Omega} = \frac{1}{\bar{r}} + \frac{\bar{r}^2}{4} (1 + 3 \cos 2 \theta), \tag{8}$$

for the value $\bar{C} = 0$ which corresponds to $C(0) = \frac{3}{2}$ by virtue of the relation

$$C(0) = \frac{3}{2} + \mu^{\frac{2}{3}} \bar{C} + O(\rho^{2/3}).$$

Applying the preceding theorem we see that we must find the conic solution of the partial differential equation

$$\frac{1}{2} \left[\left(\frac{\partial s}{\partial r}\right)^2 + \frac{1}{r^2} \left(\frac{\partial s}{\partial \theta}\right)^2 \right] - \frac{\partial s}{\partial \theta} = \frac{1}{r} + \frac{r^2}{4} (1 + 3 \cos 2\theta), \quad (9)$$

which can be expanded in powers of $\rho = \sqrt{r}$ in the vicinity of $\rho = 0$. Giving only the result, in terms of the initial $r = \mu^{1/3} \bar{r}$, we find

$$\Theta = - \left(\frac{r^3}{\mu}\right)^{7/6} \left[3 \frac{\sqrt{2}}{14} \sin 2\theta + \frac{3}{35} \frac{r^3}{\rho} \cos 2\theta + \ldots \right] (10)$$

The expression in the brackets is a series in powers of $\frac{r^3}{\mu}$, with coefficients periodic in θ. Since we do not know the radius of convergence of this series, we cannaot give a definite answer concerning the meaning of the "radius of action."

Part 2.   The deformation of the circular orbits.

This case is studied using a simplified version of Kevorkian's more general satellite problem. Kevorkian's approximation of the restricted problem uses a change of scale:

$$r = \mu^{\alpha} r'$$
$$t = \mu^{\beta} t'. \qquad (1)$$

He has shown that if r is sufficiently small one can choose $\alpha = 1/2$, $\beta = 1/4$, with the Hamiltonian

$$E' = \frac{1}{2} \left(R'^2 + \frac{\Theta'^2}{r'^2} \right) - \frac{1}{r'}$$
$$- \mu^{1/2} \frac{r'^2}{4} (1 + 3 \cos 2(\phi - \mu^{1/4} t)) \qquad (2)$$

This is the non-conservative Hamiltonian in a fixed frame,

$$\phi = (\vec{x}_1, M_2\vec{M}).$$

The moving axis $\vec{x}$ rotates with the angular velocity $\omega = \mu^{1/4}t'$, so that $\theta = \phi - \mu^{1/4} t'$.

Remarks: (1) We can write E' in the moving frame as

$$F' = \frac{1}{2} \left(R'^2 + \frac{\theta'^2}{r'^2} \right) - \mu^{1/4} \theta'$$
$$- \frac{1}{r'} - \mu^{1/2} \frac{r'^2}{4} (1 + 3 \cos 2\theta). \qquad (3)$$

(2) We can also write H (Hill's Hamiltonian) in the fixed frame

$$\bar{E} = \frac{1}{2} \left(\bar{R}^2 + \frac{\bar{\Theta}^2}{r^2} \right) - \frac{1}{r} - \frac{\bar{r}^2}{4} [1 + 3 \cos 2(\phi - t)] \qquad (4)$$

(3) Kevorkian's form is interesting because this is a perturbed Keplerian Hamiltonian $F_o + \mu F_1$, without singularity in F_1. It will now be shown that Kevorkian's form is not an approximation of Hill's Hamiltonian but rather it is its equivalent.

Note that the Keplerian part of E in Equation (4) is invariant bo Cartan's transformation

$$\bar{r} = \epsilon^{2/3} r'$$

$$t = \epsilon t'. \tag{5}$$

This gives

$$\bar{E} = \epsilon^{2/3} \left[\frac{1}{2} \left(R'^2 + \frac{\Theta'^2}{r'^2}\right) - \frac{1}{r'}\right]$$

$$- \epsilon^{4/3} \frac{r'^2}{4} [1 + 3 \cos 2(\phi - \epsilon t')]$$

or

$$\bar{E} \times \epsilon^{2/3} = \left(R'^2 + \frac{\Theta'^2}{r'^2} - \frac{1}{r'}\right.$$

$$- \epsilon^2 \frac{r'^2}{4} [1 + 3 \cos 2(\phi - \epsilon t')]. \tag{6}$$

Kevorkian's Hamiltonian is obtained for $\epsilon - \mu^{1/4}$

Note also that

$$C(\bar{E}) = \epsilon^{-2/3} C(E'), \ 0(\mu^{2/3})$$

or

$$C = \frac{3}{2} + \left(\frac{\mu}{\epsilon}\right)^{2/3} C(E') + \sigma(\mu^{2/3}). \tag{7}$$

We restrict ourselves to the case

$$C(0) \to \frac{3}{2} \quad \text{when} \quad \mu \to 0 \quad \text{or}$$

$$\epsilon = \mu^\beta, \quad 0 \leq \beta < 1 \text{ and } C(E') \text{ is finite.}$$

Proposition 2: If the distance $M_2M = r$ is of the order $\mu^{1/3} \times \mu^{2/3\beta}$, $0 \leq \beta < 1$, then Hill's approximation of the restricted problem can be written in the Keplerian perturbed form:

$$E' = \frac{1}{2} \left(R'^2 + \frac{\Theta'^2}{r'^2}\right) - \frac{1}{r'}$$

$$- \epsilon^2 \frac{r'^2}{4} [1 + 3\cos 2(\phi - \epsilon t')]. \tag{8}$$

Note 1:   E' is the non-conservative Hamiltonian in the
fixed frame, and ε is the angular velocity of the moving system.

Note 2: $\phi = \overrightarrow{M_2 x_1}, \overrightarrow{M_2 M}$

$$r = \mu^{1/3} \varepsilon^{2/3} r'$$

$$t = \varepsilon t'.$$

APPLICATION: Deformation of the circular orbits in Kevorkian's
approximation of the restricted problem.

Dropping the primes, we study the following problem:

Find the solution of the Hamiltonian

$$E = \frac{1}{2} (R^2 + \frac{\Theta^2}{r^2}) - \frac{1}{r} - \varepsilon^2 \frac{r^2}{4} (1 + 3 \cos 2\theta), \quad \theta = \phi - \varepsilon t$$

corresponding to the initial conditions for t=0

$$r = r_o \quad , \quad \phi = 0 \quad ,$$

$$R = 0 \quad , \quad \Theta = r^2 \frac{d\phi}{dt} = \pm \sqrt{r_o}.$$

For $\varepsilon=0$, this solution is the circular direct or retrograde solution of the Keplerian problem.

(A) <u>Transformation of the equations.</u>

We take ϕ as independent variable and the four unknown functions

$$t = t(\phi) \quad , \quad r = \frac{1}{s(\phi)} ,$$

$$\Theta = \frac{1}{k(\phi)} \quad \text{and} \quad R = \frac{-s(\phi)}{k(\phi)} . \tag{9}$$

The most direct way to obtain the transformed equations is to
use the differential form of Poincaré-Cartan:

$$\bar{\omega} = R \, dr + \Theta d\phi - E \, dt,$$

then express $\bar{\omega}$ with the new variables and write the Lagrange-
Euler equations for the integral $I = \int \bar{\omega} \, dt$ to obtain the transformed equations:

$$\frac{ds}{d\phi} = S, \quad \frac{dS}{d\phi} + s - k^2 = \epsilon^2 M$$

$$s^2 \frac{dt}{d\phi} = k, \quad \frac{dk}{d\phi} = \epsilon^2 M \tag{10}$$

$$M = k^2 \frac{s^{-4}}{2} [-s(1 + 3 \cos 2\theta + 3S \sin 2\theta]$$

$$N = \frac{3}{2} k^{-3} s^{-4} \sin 2\theta.$$

(Recall that $\theta = \phi - \epsilon t$.)

For $\epsilon = 0$ the general solution of Equation (10) is

$$s = k_o^2 + \lambda\cos\phi + \mu\sin\phi, \text{ and } k = k_o,$$

which gives the Keplerian ellipse:

$$\frac{1}{r} = \frac{1}{p} [1 + e \cos(\phi - \bar{\omega})]$$

with $p = \frac{1}{k_o^2}$, $e \cos\bar{\omega} = \frac{\lambda}{k_o^2}$

$$e \sin\bar{\omega} = \frac{\mu}{k_o^2}.$$

For $\epsilon \neq 0$, we wish to find the solution corresponding to

$$e = 0 \overset{\rightarrow}{\underset{\leftarrow}{}} \begin{cases} \lambda = 0 \\ \mu = 0 \end{cases}$$

and using the idea of the variation of constants, we introduce the new variables

$$s = k^2 + u,$$

$$S = v, \tag{11}$$

which gives the initial conditions

$$k = k_o$$

$$u = v = t = 0, \text{ for } \phi = 0.$$

The transformed system is

$$\frac{du}{d\phi} - v = -2k\, \varepsilon^2\, N,$$

$$\frac{dv}{d\phi} + u = \varepsilon^2\, M,$$

$$\frac{dk}{d\phi} = \varepsilon^2\, N, \tag{12}$$

$$\frac{dt}{d\phi} = \frac{k}{(k^2 + u)^2}$$

We make the a priori hypothesis that $u(0)$ and $v(0)$ remain small and expanding the expression for M and N we keep only the linear terms in u and v:

$$M = (\alpha + \beta u) + 3(\alpha + \beta u)\cos 2\theta + \gamma v \sin 2\theta,$$

$$-2kN = (a + bu)\sin 2\theta,$$

$$\alpha = -\frac{k^{-4}}{2}\ ,\ \beta = \frac{3}{2}\,k^{-6},\ \gamma = \frac{3}{2}\,k^{-6} \tag{13}$$

$$a = -3k^{-4}\ ,\ b = 12\,k^{-6}.$$

(B)   The uniformly valid asymptotic expansion.

Let us obtain the expansion $u = u^{(0)} + \varepsilon^2\, u^{(1)} + \ldots$ with the uniformly valid condition $u^{(1)}$, $v^{(1)}$, $k^{(1)}$, $t^{(1)}$ uniformly bounded in the infinite range of the argument which needs not to be ϕ but possibly

$$\psi = (1 + \lambda\varepsilon^2)\phi\ . \tag{14}$$

Here λ is a constant unknown parameter which has to be determined by the uniformly valid conditions.

System (12) splits into two:

$$\frac{du^o}{d\psi} - v^o = 0,$$

$$\frac{dv^o}{d\psi} + u^o = 0,$$

$$\frac{dk^o}{d\psi} = 0, \tag{15}$$

$$\frac{dt^o}{d\psi} = \frac{k^o}{[(k^o)^2 + (u^o)]^2}\ ,$$

$$\frac{du^1}{d\psi} - v^1 = -2k^0 N^0 - \lambda \frac{du^0}{d\psi} \; ,$$

$$\frac{dv^1}{d\psi} + u^1 = M^0 - \lambda \frac{dv^0}{d\psi} \; .$$

(16)

We omit the latter two unnecessary equations.

M^0 and N^0 means that we use the expansions of M and N with u^0, v^0 substituted for u and v.

System (15) gives

$$u^0 = \rho\cos(\psi-\alpha),$$

$$v^0 = \rho\sin(\psi-\alpha),$$

$$k^0 = k_o = \text{(the initial value)},$$

$$t^0 = k_o^{-3} \psi + h(\psi),$$

where ρ, α are constants of integration and $h(\psi)$ is a sum of trigonometric terms and therefore it is a bounded function.

System (16) gives the equation for u^1:

$$(u^1)'' + u^1 = -2k_o \frac{dN^0}{d\psi} + M^0 - \lambda[(u^0)'' + (v^0)']$$

or

$$(u^1)'' + u^1 = -2k_o \frac{dN^0}{d\psi} + M^0 - 2\lambda u^0.$$

(17)

The uniformly valid condition requires that there cannot be $\cos\psi$ or $\sin\psi$ in the second term.

Using

$$\cos\psi \; \cos2\theta = \frac{1}{2} [\cos(2\theta-\psi) + \cos(2\theta+\psi)],$$

$$\sin\psi \; \sin2\theta = \frac{1}{2} [\cos(2\theta-\psi) - \cos(2\theta+\psi)],$$

we obtain for this term the expression

$$- \frac{k_o^{-4}}{2} [1 + 15 \cos2\theta]$$

$$+ \rho\cos(2\theta-\psi)[2\lambda + 9k_o^{-6}]$$

$$+ \rho\cos(2\theta+\psi) \frac{21}{2} k_o^{-6} \; .$$

The critical term is the second, which gives the uniformly valid condition:

$$\lambda = -\frac{9}{2} k_o^{-6}. \tag{18}$$

The initial condition, $u^0 + \varepsilon^2 u^1 = 0$ for $\psi = 0$,

gives $$\rho = -\frac{11}{8} \varepsilon^2 k_o^{-4}$$

so that we obtain

$$\psi = (1 - \frac{9}{2} \varepsilon^2 k_o^{-6})\phi$$

$$s = k_o^2 [1 + \varepsilon^2 k_o^{-6}(-\frac{1}{2} - \frac{11}{8} \cos\psi + \frac{15}{8} \cos 2\theta) + 0(\varepsilon^2)] \tag{19}$$

(C)   The method of averaging.

Consider the system (12) linearized in u and v, with $k = k_o$ and find the infinitesimal transformation:

$$u = \bar{u} + \varepsilon^2 p(\bar{u},\bar{v},\phi),$$

$$v = \bar{v} + \varepsilon^2 q(\bar{u},\bar{v},\phi),$$

in such a way that equation (12) splits in the averaged system

$$\frac{d\bar{v}}{d\psi} + \bar{u} = \varepsilon^2 (\alpha + \beta_1)\bar{u},$$

$$\frac{d\bar{u}}{d\psi} - \bar{v} = 0.$$

The remaining conditions determine p and q.

We have $\beta_1 + \beta_2 = \beta$, where β_2 is given by the condition that p and q have trigonometric terms only. Therefore,

$$\beta_2 = -\frac{15}{2} k_o^{-6} \rightarrow \beta_1 = 9 k_o^{-6},$$

where $\bar{u}$ satisfies the equation

$$(\bar{u})'' + \bar{u} = (9\varepsilon^2 k_o^{-6})\bar{u} - \varepsilon^2 \bar{k}_o^{-4} \tag{20}$$

which gives the uniformly variable

$\omega\phi$,

$$\omega^2 = 1 - \epsilon^2 9 \; k_o^{-6},$$

or

$$\psi = (1 - \epsilon^2 \frac{9}{2} k_o^{-6}).$$

(D) Discussion of the results.

The only parameter is $\epsilon \; k_o^{-3}$. Returning to Hill's problem with

$$\bar{r}_o = \epsilon^{2/3} \quad r_o = \epsilon^{2/3} k_o^{-2},$$

we see that the parameter is $(\bar{r}_o)^{3/2}$.

Let the initial conditions for Hill's problem be the circular condition for the Keplerian motion with respect to the small mass M_2. Then the polar coordinates $\bar{r}, \phi$ are given by

$$\phi = (1 + \frac{9}{2} \bar{r}_o^3) \psi,$$

$$\frac{\bar{r}_o}{\bar{r}} = 1 + (\bar{r}_o)^3 \; [-\frac{1}{2} - \frac{11}{8} \cos\psi + \frac{15}{8} \cos 2\theta] \qquad (21)$$

$$\theta = \psi(1 - \bar{r}_o^{3/2}) + \text{a bounded function of } \psi.$$

The averaged solution is obtained by neglecting the fluctuations in $\cos 2\theta$:

$$\tilde{\phi} = (1 + \frac{9}{2} \bar{r}_o^3) \psi , \qquad (22)$$

$$\frac{\bar{r}_o}{\tilde{r}} = (1 - \frac{\bar{r}_o^3}{2})(1 - \frac{11}{8} \bar{r}_o^3 \cos\psi),$$

which can be rewritten in elliptical form

$$\frac{\bar{r}_o}{\bar{r}} = \frac{1}{\tilde{p}} [1 - e \cos(\phi - \bar{\omega})], \qquad (23)$$

where $\quad \tilde{p} = 1 + \frac{\bar{r}_o^3}{2} \qquad (24)$

and $\quad e = \frac{11}{8} \bar{r}_o^3 . \qquad (25)$

The longitude of the apocenter varies slowly with ϕ:

$$\bar{\omega} = (1 - \frac{9}{2} \bar{r}_o^3) \phi. \tag{26}$$

The solution is exact to order ε^2 and fluctuates around the averaged orbit with the amplitude

$$\frac{15}{8} \bar{r}_o{}^3.$$

REFERENCES

Kevorkian, J. (1962), Astronomical Journal, 67, 204.

Levi-Civita, T. (1903), C.R.Acad.Sci.Paris, 163, 221.

Nahon, F. (1978), Deformation of circular orbits, in Proc. IAU Symp. 81, (Ed: R. Duncombe), to appear.

Szebehely, V. (1967), "Theory of Orbits," Academic Press, N.Y.

STABLE AND UNSTABLE MANIFOLDS IN PLANAR TRIPLE
COLLISION

Jorg Waldvogel

ETH, Zurich, Switzerland

ABSTRACT. An adequate description of triple collision in the
planar problem of three bodies may be achieved by means of the
variables introduced by McGehee (1974). One is lead to the no-
tion of the triple collision manifold T, is of dimension 5 and
contains 10 hyperbolic equilibrium points, half of which correspond to collision. Of particular interest are the solutions
emerging from an equilibrium point of collision type along its
unstable manifold. It will be shown that these trajectories
correspond to parabolic solutions of the three-body problem.
Hence the quantitative structure of T may be investigated by
numerically integrating parabolic solutions. Based on asymptotic expansions the actual calculation of such solutions is
described.

1. <u>INTRODUCTION</u>.

In recent years progress has been made in the theory of
triple collision in the problem of three bodies. McGehee (1974,
1975), taking full advantage of the invariance properties with
respect to certain scaling transformations, achieved almost complete understanding of the one-dimensional case. The concept of
the <u>triple collision manifold</u> T introduced by McGehee proved to
be very useful. These ideas were used by Simo (1977, 1978) in
order to derive necessary conditions for the extension of triple
collision solutions by Easton's method.

The planar case was investigated by Waldvogel (1976, 1977),
and the importance of <u>parabolic solutions</u> was pointed out. Common to all those results is the fundamental instability of triple

Victor G. Szebehely (ed.), Instabilities in Dynamical Systems. 263-271.

close encounter solutions: a radically different post-history
results from a slight change in the initial conditions.

Here we shall first follow McGehee and Simo in describing
the triple collision manifold T of the planar problem of three
bodies. Then there will be shown that T contains all the parabolic solutions. Most numerical calculations on T can thus be
done with parabolic solutions. In particular, trajectories between two equilibrium points of T, needed for a possible Easton
extension, will be calculated as <u>doubly</u> parabolic solutions by
solving a two-point boundary value problem in a doubly infinite
time interval.

2.   SCALING.

Let $m_j > 0$ and $\underline{x}_j \quad R^2$ $(j = 1,2,3)$ be masses and position vectors of the three mass points in a three-body problem,
measured in a Cartesian frame of reference whose origin is at
the center of mass. The canonically conjugated momenta are

$$\underline{P}_j = m_j \underline{\overset{\circ}{x}}_j \quad (j = 1,2,3), \quad \equiv \frac{d}{dt} , \tag{1}$$

where t is the physical time. For convenience we use a compact
notation by defining

$$x = \begin{bmatrix} \underline{x}_1 \\ \underline{x}_2 \\ \underline{x}_3 \end{bmatrix}, \qquad p = \begin{bmatrix} \underline{p}_1 \\ \underline{p}_2 \\ \underline{p}_3 \end{bmatrix}, \qquad x,p \in R^6 . \tag{2}$$

With the mass matrix

$$M = \text{diag}(m_1, m_1, m_2, m_2, m_3, m_3) \tag{3}$$

the kinetic energy T and the force function V become

$$T(p) = \frac{1}{2} p^T M^{-1} p , \qquad V(x) = \sum_{j<k} \frac{m_j m_k}{\underline{x}_j - \underline{x}_k} , \tag{4}$$

and the Hamiltonian equations of motion are

$$\dot{x} = M^{-1} p, \qquad \dot{p} = f(x), \qquad f(x) = \text{grad } V(x). \tag{5}$$

The center-of-mass integrals may be written as

$$Ax = 0 , \qquad Bp = 0 , \tag{6}$$

where

$$B = \begin{bmatrix} 1 & 0 & 1 & 0 & 1 & 0 \\ 0 & 1 & 0 & 1 & 0 & 0 \end{bmatrix} , \qquad A = BM . \tag{7}$$

Furthermore, Equations (5) allow the energy integral

$$T(p) - V(x) = h = \text{const} \tag{8}$$

and the angular momentum integral

$$x^T Cp = c = \text{const} , \qquad C = \begin{bmatrix} 0 & 1 & & & & \\ -1 & 0 & & & & \\ & & 0 & 1 & & \\ & & -1 & 0 & & \\ & & & & 0 & 1 \\ & & & & -1 & 0 \end{bmatrix} \tag{9}$$

We now introduce new scaled variables $z, u \in R^6$ and s according to the homothetic transformation

$$x = rz , \qquad p = r^{-1/2}u , \qquad dt = r^{3/2}ds , \qquad ' \equiv \frac{d}{ds} , \tag{10}$$

where $r > 0$ is a scaling factor. If r is chosen as a small constant, (10) is the "blow up transformation" considered by Waldvogel (1975), which leaves the equations of motion (5) invariant. Following McGehee (1974) r will be chosen as the square root of the moment of inertia,

$$r^2 = x^T Mx ; \tag{11}$$

r is a variable quantity characterizing the size of the three-body system at every instant.

In order to transform the equations of motion (5) to the new variables first differentiate (11) with respect to t, then insert $\dot{x}$ from (5) and use (10):

$$\dot{r}r = x^T M x = x^T p = r^{1/2} z^T u \; ;$$

hence

$$r' = r(z^T u)$$

is the differential equation satisfied by $r(s)$. Transforming Equations (5) by means of (10) now yields

$$z' = -z(z^T u) + M^{-1} u$$

$$\hspace{8cm}(13)$$

$$u' = \frac{1}{2} u(z^T u) + f(z) \; ,$$

where the homogeneity relation

$$f(rz) = r^{-2} f(z)$$

satisfied by the vector function $f(z)$ was used. The integrals (6) become

$$Az = 0 \; , \quad Bu = 0 \; , \hspace{5cm}(14)$$

and Equation (11) yields the new constant of motion

$$z^T M z = 1 \; . \hspace{6cm}(15)$$

The energy and angular momentum equations are transformed into

$$T(u) - V(z) = rh \; , \hspace{5cm}(16)$$

$$r^{1/2} z^T C u = r^{1/2} \gamma = c \; . \hspace{4cm}(17)$$

For any solution of (5) not ending in triple collision we have $r > 0$. However, if a solution passes near by triple collision $r(s)$ will be small in an interval of the s-axis. On the other hand, the transformed system (12), (13) has solutions with $r(s) = 0$. Hence these solutions correspond to the behaviour of the three-body problem at triple collision; their entirety is said to form the <u>triple collision manifold</u> T (embedded in phase space).

According to Equation (16), the energy interval on T is

$$T(u) - V(z) = 0 . \tag{18}$$

The triple collision manifold T is thus defined by the differential equations (13) which allow the integrals of motion as given by Equations (14), (15) and (18). This manifold is 6 dimensional. As follows from (17), the angular momentum c vanishes identically on T, but $\lambda = z^T Cu$ is no longer a constant of motion. However, Equations (13) imply

$$\gamma' = - \frac{1}{2}(z^T u)\gamma ;$$

hence the set

$$\gamma = z^T Cu = 0 . \tag{19}$$

forms an invariant 5-dimensional submanifold $N \subset T$, referred to as the nonrotating triple collision manifold.

3. MANIFOLDS.

The equations (13) as well as the integrals defining T and N are invariant under the rotation

$$R = \begin{bmatrix} D & & 0 \\ & D & \\ 0 & & D \end{bmatrix} , \qquad D = \begin{bmatrix} \cos\phi & -\sin\phi \\ \sin\phi & \cos\phi \end{bmatrix} .$$

This degree of freedom may be eliminated from T and N leaving triple collision manifolds of dimensions 5 or 4. The structure of T is strongly influenced by the equilibrium points, i.e., the solutions with $z'(s) = u'(s) = 0$.

In the following z_e, u_e denote the values of the corresponding vectors at an equilibrium point. With the notation,

$$\beta(s) = z^T u , \qquad \beta_e = z_e^T u_e$$

they satisfy

$$M^{-1} u_e - \beta_e z_e = 0 , \qquad f(z_e) + \frac{1}{2} \beta_e u_e = 0 .$$

Multiplying the second equation by z_e^T yields

$$\beta_e = \sigma\sqrt{2V(z_e)} , \qquad \sigma = \pm 1 . \tag{20}$$

There follows

$$f(z_e) + V(z_e)Mz_e = 0 , \qquad z_e^T Mz_e = 1 , \qquad \text{and} \tag{21}$$

$$u_e = \beta_e \cdot Mz_e . \tag{22}$$

Equation (21) agrees with the condition that z_e is a central configuration with unit moment of inertia. In general there are 5 central configurations (3 collinear ones and 2 triangular ones according to the ordering of the masses). To each central configuration z_e there are 2 distinct values of u_e according to the sign σ in Equation (22). Hence T contains 10 equilibrium points; the 5 with $\sigma = -1$ correspond to collision, the others correspond to ejection. Equation (22) implies

$$\gamma_e = \beta_e z_e^T CMz_e = 0 ;$$

therefore the 10 equilibrium points (z_e, u_e) are all $\in N$.

All the equilbrium points are unstable (hyperbolic); for a detailed discussion the reader is referred to Simo (1978) and Waldvogel (1977). The dimensions of the stable and unstable submanifolds $\subset N$ at the equilibrium points are given in the following table:

	collinear		triangular	
	stable	unstable	stable	unstable
collision	1	3	2	2
ejection	3	1	2	2

A typical triple close encounter will take place as follows. As the system approaches triple collision, its trajectory in phase space will approach one of the equilibrium points of collision type, say P_1. Then, the trajectory will stay close to

N and follow a path on N emerging from P_1 along its unstable
manifold.   If this path does not pass near by any other equilibrium point, as most often will happen, the triple encounter will
end in the fast escape of one mass point.

On the other hand, if the path on N passes close to another
equilibrium point, say P_2 of ejection type, its continuation will
follow the unstable manifold of P_2, i.e., the system will eject
from another triple collision.   This is an example of the <u>exten-
sion of one particular solution</u> through triple collision.

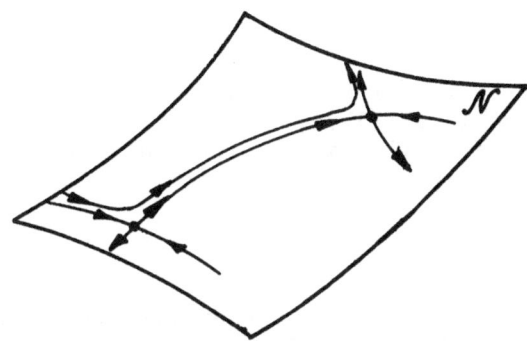

Figure 1

As Figure 1 suggests, however, the behaviour after the ejection
may change drastically due to small changes before collision.
For Easton's (1971) extension to be possible it would be required
that the behaviour after ejection from P_2 depends <u>continuously</u>
on the state before collision at P_1.

A quantitative investigation of the triple collision manifold   must be done by numerically integrating the differential equations (13).   For this purpose, however, binary collisions have to be regularized.   One possibility is to use the
global regularization suggested for the planar three-body problem by Waldvogel (1972).

If z_j, u_j are the <u>complex</u> numbers corresponding to the
2-vectors $\underline{z}_j, \underline{u}_j$ (j = 1, 2, 3)   new complex coordinates ξ, η
and complex momenta v, w are introduced by

$$z_1 = z_3 + (\xi^2 - \eta^2)^2, \qquad u_1 = \frac{1}{4} \cdot \frac{v\bar{\eta} - w\bar{\xi}}{2\bar{\xi}\bar{\eta}(\bar{\xi}^2 - \bar{\eta}^2)} \quad ,$$

$$z_2 = z_3 + (\xi^2 + \eta^2)^2, \qquad u_2 = \frac{1}{4} \cdot \frac{v\bar{\eta} + w\bar{\xi}}{2\bar{\xi}\bar{\eta}(\bar{\xi}^2 + \bar{\eta}^2)} \quad ,$$

$$\hspace{10cm} (23)$$

$$z_2 - z_1 = (2\xi\eta)^2, \quad u_3 = -u_1 - u_2 = \frac{1}{4} \cdot \frac{-v\bar{\xi} + w\bar{\eta}}{(\bar{\xi}^2 - \bar{\eta}^2)(\bar{\xi}^2 + \bar{\eta}^2)} \cdot$$

Since the center-of-mass integrals may be eliminated by means
of the transmformations (23), the new differential equations form
a system of 8th order allowing the integrals (15), (18) and the
invariant manifold (19).

4. PARABOLIC SOLUTIONS.

For settling the question whether some cases of planar
triple collision can be extended by Easton's method, trajectories
connecting an equilibrium of collision type with one of ejection
type play a crucial. Therefore, it is useful to establish a
direct relationship between trajectories on the triple collision
manifold T and solutions of the three-body problem.

THEOREM: The nonrotating triple collision manifold N of a
three-body problem is equivalent to the manifold of all solutions
of the original problem with vanishing energy and vanishing angular momentum (h = c = 0).

Proof: For a solution with h = c = 0 the scaling transformation (10) with r > 0 produces Equations (13), which are
independent of r, as are the integrals (15), (18) and (19).
These are exactly the equations defining N. Assume, on the other
hand, a trajectory $C N$ to be given. Since the integrals (15),
(18), and (19) are a consequence of the differential equations
(13) a function r(s) > 0 can always be supplemented by means
of Equation (12).

A trajectory emerging from an equilibrium point of collision
type or ending in an equilibrium of ejection type will be called
a parabolic solution. The corresponding solution of Equations
(5) is characterized by $x \rightarrow \infty$, $p \rightarrow 0$ as the equilibrium is
approached. A trajectory connecting two equilibria will be called
a doubly parabolic solution.

The problem of calculating doubly parabolic solutions is difficult. Whether they are calculated in physical space or on N, one must work in doubly infinite intervals, $-\infty < t < \infty$ or $-\infty < s < \infty$. Mapping these into finite intervals causes singularities at the end points.

One possibility is to consider a family of parabolic solutions, where initial conditions at $t = -T$ (a sufficiently large negative value) are taken from the expansion corresponding to Siegel's series for triple collision (see Hulkower, L978). For each member of the family the index of the excaping body (as $t \rightarrow \infty$) is determined. Finally, a bisection type method allows one to iterate on the case where no unique escaper can be found. This is the doubly parabolic solution. An example in the equal mass case ($m_1 = m_2 = m_3$) was given by Waldvogel (1976).

5.   REFERENCES.

Easton, R. (1971), J. Diff. Equ. 10, pp. 92-99.

Hulkower, N.E. (1978), "The Zero Energy Three Body Problem", Indiana Univ. Math. J. 27, pp. 409-447.

McGehee, R. (1974), Invent. Math. 27, pp. 191-227.

Simó, C. (1977), Actas II Asamb. Astron. y Astrofís, Cádiz.

Simó, C. (1978), "Masas para la regularización de la colisión triple", Actas Jorn. Mat. Luso-Espanholas, Aveiro.

Waldvogel, J. (1977), Bull. Acad. Roy. Belg., Classe des Sci. 63, pp. 34-50.

Waldvogel, J. (1976), Cel. Mech. 14, pp. 287-300.

Waldvogel, J. (1975), Cel. Mech. 11, pp. 429-432.

Waldvogel, J. (1972), Cel. Mech. 6, pp. 221-231.

PART V

ABSTRACTS OF SEMINAR-CONTRIBUTIONS

CLUSTER FORMATION AND GRAVITATIONAL INSTABILITIES

S. J. Aarseth

Institute of Astronomy, Cambridge, England

ABSTRACT. N-body simulations are used to model the clustering of galaxies in an expanding universe. Initial fluctuations in density grow with time and the final distributions show large bound clusters. Several calculations are made for systems with 1000 and 4000 galaxies and the results show good agreement with the observed distributions of galaxies within the local super-cluster. The effectiveness of the gravitational instabilities is illustrated by time-sequence movies which include rotational displays to give the viewer a sense of the three-dimensional system. This work will be reported fully in a series of papers, to be published in the Astrophysical Journal together with Drs. J. R. Gott and E. L. Turner.

APPLICATION OF A NEW NUMERICAL METHOD FOR THE INTEGRATION OF THE EQUATIONS OF MOTION FOR AN N-BODY SYSTEM

A. Carusi

Laboratorio di Astrofisica Spaziale, CNR, Frascati, Italy

ABSTRACT. A new numerical method for the integration of the equations of motion of a gravitating system of bodies has been developed by Greenspan (1973) and by LaBudde and Greenspan (1973, 1976a, 1976b).

The most remarkable aspect of the method is the conservation of the invariants of motion, in the same way as in classical mechanics. The method consists of recursive formulas for the computation of the coordinates and velocity components of each body. The computation of these quantities is made at each time step, taking into account the mutual gravitational forces of all bodies.

The precision of the method depends on the accuracy of the initial data of the planetary positions and velocities. Because of the recursive formulas, the final error depends upon the choice of the computational time step Δt. The error is proportional to the second and third power of Δt respectively, for the velocities and positions.

The difference between this method and others (Taylor series, Runge Kutta, etc...) is essentially a different formulation of the gravitational force and potential. Generally, the gravitational force is defined in the same manner as in classical mechanics with the consequence that the invariants of motion are no longer constant.

Greenspan's method gives a different solution to this problem using an approximate formula for the force, and an exact expression for the potential, in such a way that the invariants

of the motion are rigorously conserved at each time step. This
approximation introduces an error that is essentially equal to
the admitted tolerance in the convergence of Newton's method
for the solution of the equations of motion. The two great
advantages of Greenspan's approach are: i) the solution has a
greater precision, and (ii) the computing time is short, because
only simple algebraic equations are to be solved.

    After some tests, Greenspan's method has been applied to
many dynamical problems. Firstly, it was used to compute the
past and future orbit of a fireball, that approached the Earth
in 1972, passing through the atmosphere at a height of about
50 km. This close approach resulted in a slight change in the
orbital parameters of the fireball (Carusi and Massaro, 1976).
The computation showed that Greenspan's method represents a
powerful approach for those problems in which close interactions
between two bodies are considered. Starting from these considerations, a great number of close encounters between Jupiter
and fictitious minor bodies have been computed, in order to
study the effect of a single close encounter on the orbital evolution of a minor body (Carusi and Pozzi, 1978b). Frequently
temporary satellite-capture has been found, suggesting a continuation of research on these events (Carusi, et al., 1978).
The orbits of one hundred bodies have been studied with initial
conditions nearly tangent to Jupiter's orbit. Temporary satellite-
capture was found in 56 percent of the cases and about 60 percent of the orbits changed their aphelion with their perihelion,
or vice-versa. A similar behaviour provides a mechanism to
transform long-period comets into short-period ones.

REFERENCES.

Carusi, A. and Pozzi, F. (1978a), "The Moon and The Planets",
    in press.

Carusi, A. and Pozzi, F. (1978b), "The Moon and The Planets",
    in press.

Carusi, A., Pozzi, F. and Valsecchi, G. (1978), Proc. of IAU
    Symp. 81, R. Duncomb (ed.), Tokyo, Japan.

Carusi, A. and Massaro, E. (1976), Astrophys. Letters 17, p.
    113.

Greenspan, D. (1973), "Discrete Models", Addison-Wesley, Reading,
    Mass.

LaBudde, R.A. and Greenspan, D. (1976a), Numer. Math. 25, p. 323.

LaBudde, R.A. and Greenspan, D. (1976b), Numer. Math. 26, p. 1.

LaBudde, R.A. and Greenspan, D. (1973), "Discrete Models",
    Addison-Wesley, Reading, Mass.

EVOLUTION OF A MISSION CONCEPT FROM A SOLAR PROBE TO A SOLAR ORBITER

Professor G. Colombo

University of Padua, Padua, Italy

ABSTRACT. In early 1976, an alternative option to the dual probe out-of-ecliptic (O.O.E) mission via Jupiter swing-by was suggested (G. Colombo, D. Lautman, and G. Pettengill, 1976). In the mission plan of O.O.E. (now called Solar Probe Orbiters) two Pioneer type spacecraft (assembled in a single package) are sent to Jupiter. Before the fly-by's closest approach to the giant planet, the two spacecrafts are separated and are injected, by proper mid-course correction, in two paths having different impact parameters with Jupiter. The gravitational effect of Jupiter changes drastically the orbits of the two spacecraft. After encounter with Jupiter the two spacecrafts move in two orbits normal to the plane of the ecliptic in opposite directions (one going south, the other north).

Considering the high efficiency of the Jupiter gravity assist, we suggested then to combine the solar polar mission with a solar probe mission, using the gravity pull of Jupiter for injecting one of the two probes in an orbit impacting the sun or having a perihelion distance of a few solar radii, (Anderson, Colombo, Friedman and Lau, 1976).

A probe at a distance as close as 4 solar radii to the sun, and possibly closer, could conduct particle and field experiments that are truly unique and could determine the quadrupole moment of the sun with the accuracy of a few parts in 10^8 or better. The main technical problems posed by the severe environment and the required measurement accuracy (thermal shields, communications, attitude control, compensation of non gravitational forces, etc.) were found to be solvable with the present technological capabilities (Randolph, 1978).

A feasibility study of a mission where the spacecraft is injected on an orbit about the sun with a period less than two
years, possibly one year, keeping the perihelion distance less
than 4 solar radii, is underway.

The main experiments and measurements will be repeated
several times at any perihelion passages. The scientific lifetime of the artificial planet is limited by the deterioration
of the thermal shield.

Because after Jupiter fly-by, the spacecraft will be in an
orbit with a semimajor axis larger than 2.61 a.u. corresponding
to a period of roughly 4 years, one should think of a method for
reducing the orbital energy after encounter with Jupiter. The
most trivial solution of this problem is obtained by retro-
firing a rocket at or close to perihelion. A change of velocity
of the order of 1 to 2 km/sec will inject the spacecraft in an
orbit with a period of 1 to 2 years. There are several technical and scientific reasons, however, for considering other methods.
The use of radiation pressure in order to decelerate the spacecraft as it goes from 1 a.u. to the sun is being studied. A
preliminary computation shows that a sail (or "parachute") of
reasonable dimensions (100 ∿ 200 m diameter), deployed at 1 a.u.
from the sun and jettisoned near the sun (at 0.1 ∿ 0.2 a.u.),
will reduce the period of the orbit to 1.5 years.

REFERENCES.

Anderson, J., Colombo, G., Friedman, L. and Lau, E. (1976), "An
    Arrow to the Sun", International Symp. on Experimental
    Gravitation, Pavia, Italy.

Colombo, G., Lautman, D. and Pettingill, G. (1976), "An Alter-
    native Option to the Dual-Probe Out-of Eliptic Mission via
    Jupiter Swingby", NASA TM-X-71097, p. 37-47.

Randolph, J.E. (1978), "Solar Probe Study", Jet Propulsion Labora-
    tory Publication 78-70.

SIMULATION OF DYNAMICAL AND THERMAL BEHAVIOUR OF SKYHOOK SYSTEM

Professor G. Colombo

University of Padua, Padua, Italy

ABSTRACT. Under contract NAS 8-32199, the Smithsonian Astrophysical Observatory has developed a complex software package
for the simulation of the dynamical and thermal behavior of a
Tethered Satellite System (Colombo, et al, 1974 and 1976). The
approach embodied in this simulation has been to discretize the
tether into a set of N point elements each with the physical
characteristics (mass, area, conductivity, etc.) of a finite
length tether segment. These point elements are employed as the
centers of application of the externally applied forces and the
internal tether tension, as well as the centers of accumulation
of the heating and cooling thermal fluxes. A system of coupled
differential equations have been derived which represent both
the dynamical and thermal variation of each element. The dynamical evolution of the system is obtained by numerically integrating these equations. The cartesian coordinates of each element are tabulated as a function of time and the displacement,
tension, and temperature of each element is displayed. The displays of the dynamical variables are presented as projections of
the coordinates onto an orbital reference system defined by the
motion of one particular element selected by the user.

The unique features of this simulation system are its
generality, the extensive force model employed, and the inclusion
of thermal effects. In particular, the dynamical forces include
(1) a terrestrial gravity field expansion composed of spherical
hormonics to degree and order 20, (2) a full earth gravity anomaly
model represented by a surface layer density distribution of arbitrary complexity, (3) solar and lunar gravitational forces,
(4) atmospheric drag forces with realistic density models, (5)
solar radiation pressure, (6) earthshine radiation pressure and

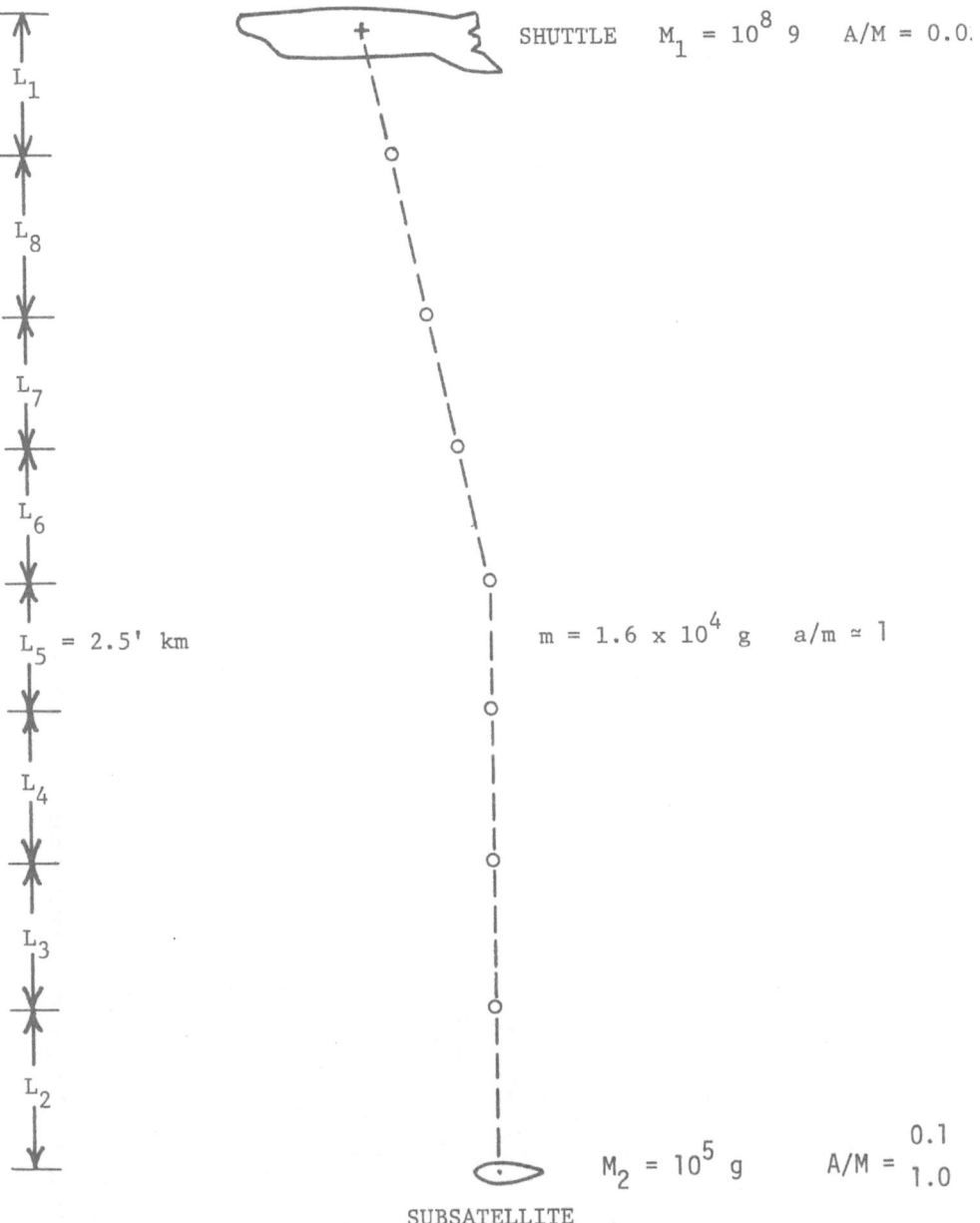

Fig. 1 - Nine-mass-point representation of 20 km tether.

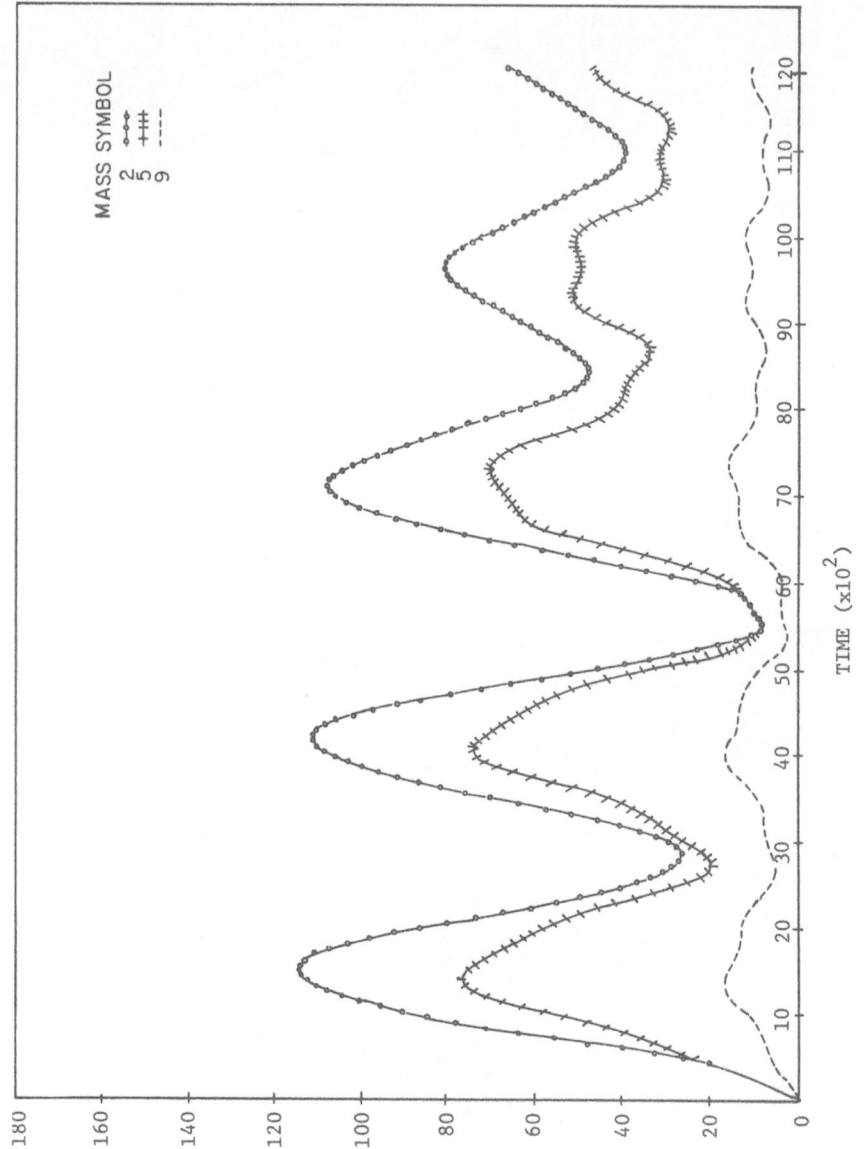

Fig. 2 – Example of "in-plane" motion of each mass point of the configuration in Fig. 1.

(7) tidal forces. The tether mechanical force model includes:
(1) a linear stress strain relationship modified by the thermal
expansion characteristics of the tether material, (2) a visco-
elastic damping force and (3) an arbitrary control law capability
for deployment, retrieval and station keeping modes of operation.
The present thermal model includes heating from solar radiation,
earthshine radiation, and atmospheric drag and cooling by radiation.

Up to 20 elements may be employed in the current software
configuration. There are no restrictions on the mechanical,
thermal or bulk characteristics of any tether segment or on the
initial conditions representing the eccentricity or orientation
of the Shuttle orbit.

Figure 1 illustrates the general configuration for the use
of a 20 km tether linking the shuttle and subsatellite with the
tether represented by 9 point elements. Figure 2 is a sample
of the graphical display of the "in-Plane" motion of each mass
point other than the Shuttle for a period of approximately two
orbits. The vertical scale is in cm and the horizontal scale is
in seconds. Symbol "2" represents the subsatellite motion relative to the Shuttle. All displacements are referenced to their
value at t = 0.0 so that the motion of all elements could be
displayed on the same scale. In actual fact, each point is
separated by 2.5 km = 250,000 cm from its nearest neighbors.

REFERENCES.

Colombo, G., Arnold, D.A., Binsack, J.H., Gay, R.H., Grossi,
     M.D., Lautman, D.A. and Orringer, O.(1976), "Dumbbell
     Gravity-Gradient Sensor: A New Application of Orbiting
     Long Tethers", Smithsonian Astrophys. Obs. Reports in
     Geoastronomy No. 2.

Colombo, G., Gaposchkin, E.M., Grossi, M.D. and Weiffenbach, G.C.
     (1974), "Shuttle-Borne 'Skyhook': A New Tool for Low-
     Orbital-Altitude Research", Smithsonian Astrophys. Obs.
     Reports in Geoastronomy No. 1.

THE JACOBIAN INTEGRAL IN THE ELLIPTIC RESTRICTED THREE-BODY PROBLEM

R. Dvorak

Astronomisches Institut, Universität Graz, Austria

ABSTRACT. The Jacobian Integral in the circular restricted three body problem is a simple analytic expression (Szebehely and Giacaglia, 1964). The more complex elliptic problem has no such integral, but nevertheless we were able to construct a series expansion to the powers of the eccentricity e up to the third order. Some specific terms are developed to higher orders and it seems that there appear secular terms in the Jacobian Integral for the elliptic case (Delva and Dvorak, 1978). Although no final answer on the secular behaviour can be given some specific properties of this "integral" may be obtained.

REFERENCES.

Delva, M. and Dvorak, R. (1978), Astron. Astroph., to appear.

Szebehely, V. and Giacaglia, E.O. (1964), Astron. J. 69, p. 230.

EFFECT OF MASS-PARAMETER ON THE PERIODIC ORBITS OF THE RESTRICTED PROBLEM OF THREE BODIES

G. Gómez

Seccio de Matematiques, Universitat Autonoma de
Barcelona, Spain

ABSTRACT. Families (a) and (b) of the restricted problem of
three bodies are symmetric periodic orbits generated, by analytic continuation, from retrograde infinitesimal orbits around
the collinear equilibrium points, L_3 and L_1 respectively.

For certain values of the mass parameter these two families
are well known ($\mu = 0.5$ by Henon; $\mu = 0.012$, by Broucke). There
are also some results from Bartlett and Wagner for $\mu = 0.5$,
0.9064, 0.9085, 0.9091, 0.9650.

With a predictor-corrector algorithm based on the integration of the first variational equations we obtained the evolution of family (a) for the following values of μ: $\mu = 0.01$,
0.02 (0.02) 0.08 (0.06) 0.92 (0.02) 0.98, 0.99.

By symmetry considerations this variation of the mass-
parameter includes also the evolution of family (b).

The characteristic curve obtained for these values of μ
is a continuous deformation of the one that we have for the
Copenhagen Category ($\mu = 0.5$). There is good agreement with the
above results, except with the ones of Bartlett and Wagner for
large values of μ.

Our results extrapolated to $\mu = 0$ also agree with Hill's.

Using the RKF78 routine with local error of 10^{-15}, 10^{-13} ensures a good conservation of Jacobi's constant (to 10^{-11}). Syn-
odical coordinates (x,y) are used except when regularization makes
the use of Thiele's variables more convenient. Periodicity

conditions guarantee that $|\dot{x}| < 10^{-8}$ if $|y| < 10^{-8}$.

Linear stability is studied by analyzing the variational matrix at the half period. When changes of stability are observed, possible bifurcations are looked for by studying Poincare's map.

To generate families, initial conditions for an infinitesimal periodic orbit around L_3 are needed. This is obtained analytically by linearizing the equations near this point. When the orbit is found, we follow the family in the (x,c) space (c = Jacobi's constant) using the arc-parameter, $(\Delta x)^2 + (\Delta c)^2$. The value of this parameter is conveniently modified depending on the curvature of the characteristic curve and the rate of variation of the parameter of stability.

SEMI-ANALYTICAL LUNAR EPHEMERIS

J. Henrard, M. Moons and D. Standaert

Department of Mathematics, Facultes Universitaires de Namur, Belgium

ABSTRACT. A theory of the motion of the Moon is being developed. The final aim is to predict the position of the center of the Moon and its libration around this center with the accuracy of $\pm$ 10 cm.

The method used to develop this theory is semi-analytical. The techniques and algorithms used are those of analytical theories but the expansions are done around nominal values of the small parameters rather than around zero and numerical truncation of the expressions becomes an important tool.

The Main Problem of the Lunar Theory and the perturbations due to the oblateness of the Earth (terms in J_2, J_3, J_4) have been computed within the required accuracy. Work is being done on two problems: the planetary perturbations and the libration of the Moon.

The planetary perturbation being investigated is the direct effect of Venus (considered as being on a mean Keplerian orbit). Very long period terms appear in these perturbations and a decision has to be made whether to keep them as periodic terms or to expand them in powers of t. The accuracy and the period of validity of the theory depend on this decision. We plan to answer this dilemma by two closely connected theories: one would be valid for a period of a few thousand years but with an accuracy of only a few meters while the other, more precise, would be valid for only 100 years.

The theory of the libration of the Moon is based upon Andoyer's variables and Hamiltonian perturbation theory. The expressions are expanded around a mean equilibrium corresponding to Cassini's empirical laws. We find that the theory is very sensitive to the expressions used in describing the motion of the Moon around the Earth.

ON THE ORIGIN OF THE KIRKWOOD GAPS

T. A. Heppenheimer

Center for Space Science
Fountain Valley, California 92708

ABSTRACT. It is proposed that the Kirkwood gaps are primordial, representing regions where asteroids failed to form by accretion. A brief scenario is presented to indicate the main features of a model for the early history of the asteroids. An analytical treatment is given for the effects of a solar nebula upon the eccentricity-pumping of asteroids, due to secular perturbations and commensurability-type resonances associated with Jupiter. It is shown that nebular effects promote growth of main-belt asteroids; but in commensurability regions, growth is inhibited. A discussion is given of two related problems: the origin of asteroidal eccentricities and inclinations, and the likelihood that Jupiter suffered major changes in its semimajor axis during its formation. It is suggested that in view of these problems, the present theory should be regarded as illustrative of viewpoints which in time may yield a correct theory.

GENERATING ORBITS FOR STABLE CLOSE ENCOUNTER PERIODIC
SOLUTIONS OF THE RESTRICTED PROBLEM

D. L. Hitzl

Lockheed Research Laboratory, Palo Alto, California

ABSTRACT. The second species periodic solutions of the restricted three-body problem have been investigated in the limiting
case of $\mu=0$. The term second species solution is due to Poincaré
and is basically a Keplerian elliptical orbit about the primary
mass which, at some time, makes a close approach to the secondary mass. The new result is that for $\mu=0$ second species orbits
are critical (i.e., are located on the stability boundary) if,
and only if, the Jacobian constant C has an extremum. These
limiting orbits are of great interest since, for small $\mu>0$,
"neighboring" orbits will have a finite (but small) region of
stability.

REFERENCES

Hitzl, D.L. and Hénon, M. (1977), Celestial Mechanics 15, p. 421.

Hitzl, D.L. (1977), Journal of the Am. Inst. of Aeronautics and
    Astronautics 15, p. 1410.

_____ and _____ (1977), Acta Astronautica 4, p. 1019.

OPTIMAL CONTINUOUS TORQUE ATTITUDE MANEUVERS

John L. Junkins and James D. Turner
Virginia Polytechnic Institute and State University
Blacksburg, Virginia

ABSTRACT. A non-singular formulation of the necessary conditions
for optimal large angle rotational maneuvers is presented. The
formulation is for the case of a rigid asymmetric vehicle having
a control system capable of generating a continuous torque history; it is valid for an arbitrary specification of terminal orientation and angular velocity states. Analytical solutions of
the two-point-boundary-value problem are extracted for single
axis maneuvers, and a relaxation process is proposed for reliable
numerical solutions for arbitrary boundary conditions. The
validity of the necessary condition formulation is supported by
analytical solutions for special cases; the relaxation process
for general boundary conditions is illustrated by numerical
examples.

The complete paper is available as an AIAA/AAS Astrodynamics
Conference (1978) preprint.

FINAL EVOLUTIONS OF PARABOLIC-HYPERBOLIC TYPE
IN THE RESTRICTED THREE-BODY PROBLEM

J. Llibre
Universitat Autonoma de Barcelona, Spain

ABSTRACT. Following Chazy we call "final evolution" of the restricted three-body problem the behavior of the infinitesimal
body when time increases to $\pm \infty$.

Of the eight types of final evolutions of the problem (P^+,
P^-, parabolic; H^+, H^-, hyperbolic; OS^+, OS^-, oscillatory; L^+,
L^-, Lagrange stable) we consider the following four:

- Parabolic evolutions, $P^+(P^-)$, in which the infinitesimal
body reaches infinity with zero radial velocity as the time tends
to $\pm \infty$.

- Hyperbolic evolutions, $H^+(H^-)$, in which the infinitesimal
body reaches infinity with positive radial velocity as the time
tends to $\pm \infty$.

When we consider two final evolutions associated with one
orbit, there are sixteen possibilities. For example, $H^- \cap P^+$,
means that when $t \to -\infty$ the evolution is of the hyperbolic type
and when $t \to +\infty$ it is parabolic.

It is well known that when the Jacobian constant is fixed,
the dimensionality of the space of orbits is two.

The purpose of this communication is to present the following result.

Theorem: For a fixed value of the Jacobian constant and when the
mass-parameter μ is sufficiently small, the set of orbits with
final evolution of the $P^- \cap H^+$ and $H^- \cap P^+$ types is homeomorphic
to a countable union of disjoint open intervals.

ASYMPTOTIC ORBITS AND INSTABILITY ZONES IN THE SIMPLEST MODEL OF A GALACTIC POTENTIAL.

P. Magnenat

Geneva Observatory, Switzerland

ABSTRACT. The relative importance of the third-order terms ε and η in the potential field

$$V = \frac{1}{2} (Ax^2 + By^2) - \varepsilon xy^2 - \eta x^3$$

is numerically explained. Particularly the stability of periodic orbits and the appearance of "wild" behaviour on the surface of section is studied for various values of ε and η. These experiments allow to display some interesting features of the dynamical problem with two degrees of freedom:

- Varying the coefficients of the third order terms, we obtain nine distinct topological types of surfaces of section, including those of Contopoulos and Moutsoulas (1965) and of Hénon and Heiles (1964).

- For a given ε (large enough), the area of "wildness" is enlarged for a negative value of η and shrinked for positive η, compared to the case when $\eta=0$.

- The shapes of some asymptotic curves, or separatrices, calculated in detail, show well the oscillations expected by Poincare. Besies the well known classical types (homoclinic and heteroclinic) of curves, these experiments show special behaviour of heteroclinic orbits which seem to have not been described until now. The "classical" case of a heteroclinic curve between two unstable periodic orbits occurs only if the instability parameters of both orbits are equal.

- The shapes of these curves allow us to explain the progressive invasion of the plane of section by "ergodicity".

- A useful stochasticity parameter, based on the criterion of
  Benettin, et al. (1976), is introduced whose variation reflects
  the topology of the surface of section well.

REFERENCES.

Contopoulos, G. and Moutsoulas, M. (1965), <u>Astron. J.</u> <u>70</u>, p. 817.

Hénon, M. and Heiles, C. (1964), <u>Astron. J.</u> <u>69</u>, p. 73.

GENERATION OF PERIODIC ORBITS IN THE THREE-BODY PROBLEM

J. Martinez

Catedra Especial de Tecnologias del Espacio
Universidad Politecnica de Barcelona, Spain

ABSTRACT. Relative planar periodic orbits of the three-body problem are computed as numerical continuations of absolute periodic orbits of the elliptic restricted problem. We use:

- The symmetry theorem: "If both particles m_1 and m_2 cross a fixed axis, passing through m_o (at the origin), at the same time and at right angles, the orbits of m_1 and m_2 are symmetric with respect to this axis". (Broucke, 1975).

- The periodicity criterion: "If a solution has two consecutive symmetric intersections at time $t=0$ and $t=T/2$ the solution is relative periodic with period T and the rotation angle is twice the angle between the two symmetry axes".

We take the initial conditions corresponding to the elliptic restricted problem using two-body approximation (Broucke, 1969). Heliocentric Jacobian coordinates are used. The period T and the rotation angle ϕ are $2K\pi$ where K is an integer. All symmetric periodic orbits in the elliptic restricted problem are periodic in inertial coordinates, therefore $\phi=0$ or $\phi=2K\pi$. A Runge-Kutta-Fehlberg 7/8 numerical integrator (NASA TR R287) with KS regularization is used (Bettis and Szebehely, 1971).

We consider $m_o=m_1=\frac{1}{2}-\varepsilon$, $m_2=2\varepsilon$ with the initial conditions:

$$x_{10} \; , \; y_{10}=0 \; , \; \dot{x}_{10}=0 \; , \; \dot{y}_{10} \; , \; x_{20} \; , \; y_{20}=0 \; , \; \dot{x}_{20}=0 \; , \; \dot{y}_{20}.$$

One parameter, x_1 is kept fixed to keep the scale of the solution constant. The other three parameters must verify the equations:

$$\bar{F}_1 = (\bar{r}_1 - \bar{r}_o) \times (\bar{r}_2 - \bar{r}_o) = \bar{0}, \quad F_2 = (\bar{r}_1 - \bar{r}_o) \cdot (\dot{\bar{r}}_1 - \dot{\bar{r}}_o) = 0,$$

$$F_3 = (\bar{r}_2 - \bar{r}_o) \cdot (\dot{\bar{r}}_2 - \dot{\bar{r}}_o) = 0.$$

Also the value of $\dot{y}_{10}$ is kept constant. In the first approximation the value of ε is small enough to assure convergence, (of the order of 10^{-3} or 10^{-4}). For the other members of the family we consider:

$$m_{o \ new} = m_{o \ old} - \varepsilon, \quad m_{1 \ new} = m_{1 \ old} - \varepsilon, \quad m_{2 \ new} = m_{2 \ old} + 2\varepsilon.$$

The criterion of periodicity was $\sqrt{F_2{}^2 + F_3{}^2} < 10^{-9}$.

Relative symmetric periodic orbits for the general planar three-body problem were obtained using as initial conditions the periodic orbits of the elliptic restricted problem, showing the results in a convenient rotating frame of reference (Hadjidemetriou, 1975). A complete analysis for given masses of the bodies has been made. In some cases these periodic solutions consist of double Keplerian motion where a pair of bodies moves in approximate Keplerian motion around each other and the third body moves around the binary also in an approximate Keplerian orbit. The computations were carried out on a CDC 6600/6400 during the author's stay in the Department of Aerospace Engineering at the University of Texas at Austin.

REFERENCES.

Bettis, D. and Szebehely, V. (1971), Astrophysics and Space Science 14, p. 133.

Broucke, R. (1975), Cele. Mech. 12, p. 4.

Broucke, R. (1969), Jet Propulsion Laboratory, NASA TR-1360.

Hadjidemetriou, J. (1975), Cele. Mech. 12, p. 255.

MEASUREMENT OF THE GRAVITATIONAL CONSTANT IN SPACE.

[1] P. Farinella, [2] A. Milani and A. Nobili

[1] Osservatorio Astronomico di Brera, Italy
[2] Universita di Pisa, Italy

ABSTRACT. The gravitational constant G is a universal constant known with few significant digits. The best measurements yield $G = (6.673 \pm 0.003) \times 10^{-8}$ cm^{3}s^{-2}g^{-1}, where the error estimates are uncertain. There are fundamental limitations to any measurement of small forces on the Earth because of seismic noise (as analyzed by Braginski). On the other hand, no such restrictive limitataions are present in space, where the determination of G (in astronomical units) is performed with great precision by methods of celestial mechanics.

Therefore, we propose to measure G by putting in a low Earth orbit a known mass of high density and measuring the orbit of a satellite mass under the gravitational influence of the principal mass (and of the other relevant bodies). The Earth's gravity gradient cannot be treated as a small perturbation, hence a three-body model must be used to qualitatively predict the behaviour of the satellite mass, and a numerical integration with iterative differential corrections is necessary to extract the result concerning G from experimental data.

By Hill's stability criterion it can be shown that no stable orbit near the big mass can be obtained in a low Earth orbit. On the other hand, in order to accurately determine G one must keep the satellite mass for as long as possible as close as possible to the big mass. The cumulative displacement in time t of the satellite mass under the effect of the admitted error for the acceleration is $c \, t^2 \, r^3 \, \ell^{-2}$, where r is the radius of the big mass, ℓ is the average distance of the satellite mass from the center of the big mass and c is a coefficient of the order of 10^{-11} in CGS units.

296

Therefore, it is proposed to use the "practical stability" of the equilibrium points (in the rotating reference frame) by putting the satellite mass as close as possible to the stable manifold of an equilibrium point. This could give a duration of the experiment of the order of a few thousand seconds if a very precise "initial condition gun" is used (based for instance on radiation pressure with interferometric control). This time would be enough to measure the fifth significant digit of G using a photographic film with a position resolution of a few microns.

During the motion of the satellite the big mass must be free-falling inside a spacecraft or balloon that shields all non-gravitational forces as atmospheric drag, radiation pressure, ect. In order to keep this state of drag-free motion of the masses, they must be carefully positioned inside the spacecraft to prevent the gravitational gradient from pulling them against the laboratory walls, or the spacecraft or balloon must be actively maneouvred to avoid such a collision.

HILL'S STABILITY APPLIED TO THE ASTEROIDAL BELT

[1] P. Farinella and [2] A. Nobili

[1]
[2] Osservatorio Astronomico di Brera, Italy
Universita di Pisa, Italy

ABSTRACT. We study the stability of asteroidal orbits within the framework of the restricted three-body problem (Sun, Jupiter, asteroid), assuming an initial distribution of circular orbits around the Sun and analyzing the zero-velocity curves which bound the motion of asteroids in the rotating reference frame. For each orbit, we compare the Jacobian constant J with the critical value J_{cr} which allows the escape of the body beyond L_2 , and we deduce that instability is reached for initial radii larger than about 4.2 A.U.. This result is in agreement both with the observational fact that no asteroid is found between Thule's group and Jupiter, and with the numerical results by Lecar and Franklin (1973), which show a rapid ejection of asteroids beyond 4.0 A.U. whether or not Jupiter's eccentricity is taken into account. Moreover, from an analysis of the size of the zero-velocity curves we derive the maximum eccentricity available to each asteroid, and we show that this treatment yields a qualitative agreement with the observed eccentricity distribution.

Finally, we study the effects of a random collisional process within the asteroidal belt, concluding that it could have been responsible for the drastic depopulation of the zone between 3.3 and 4.0 A.U., in which the stability parameter $(J-J_{cr})/J$ is much smaller than unity. (We note that the depopulation cannot be explained by the numerical collisionless calculations). By means of an estimation of the number of collisions needed in order to deplete this zone, we conclude that the original mass of the asteroidal belt was of the order of a small planetary mass.

REFERENCES.

Lecar, M. and Franklin, J. (1973), Icarus 20, p. 422.

NUMERICAL STUDY OF THE RECTILINEAR ISOSCELES RESTRICTED PROBLEM.

F. Puel

Observatoire, Universite de Besancon, France

ABSTRACT. We study the orbits of a particle in the plane of symmetry of two primaries of equal mass performing rectilinear Keplerian oscillation.

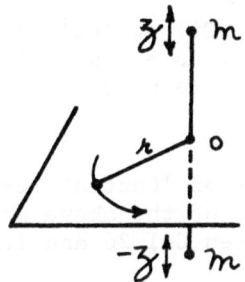

After normalization, and choosing the eccentric anomaly u of the primaries as the independent variable, the equations of motion become

$$
\begin{cases}
\dfrac{dr}{du} = \sqrt{8}\,(1 - \cos u)\dot{r} \\[2ex]
\dfrac{d\dot{r}}{du} = \sqrt{8}\,(1 - \cos u)\left\{ \dfrac{-r}{[z^2+(1-\cos u)^2]^{3/2}} + \dfrac{c^2}{r^3} \right\} \\[2ex]
\dfrac{d\theta}{du} = \sqrt{8}\,(1 - \cos u)\,\dfrac{C}{z^2} \quad .
\end{cases}
$$

Ignoring the orientation of the system, as usual in the n-body problem, we can neglect the third equation.

We have two exact solutions:

for $(r=0, \dot{r}=0)$ we have the Lagrangian point L_1 and

for $(r = \sqrt{3} z, \dot{r} = \sqrt{3} \dot{z})$ we have Lagrange's equilateral solution, L_4.

To help in the numerical search it is possible to start from circular solutions which hold for very large C. The Keplerian approximation with a force function $1/r$ instead $[r^2+z^2(t)]^{-1/2}$ or, even better, the two fixed force centers approximation $(r^2+2.5)^{-1/2}$ may be used. (In the latter 2.5 is the mean value of z^2). These provide approximations for limites of bounded orbits, for quasi-circular solutions and also for periodic solutions, i.e., solutions where the radial oscillations of the particle are commensurable with the motions of the primaries.

For the numerical research we use the method of Henon (1965, 1966). The system is explored by the method of 'Surface of Section' observing the state of the particle in the $(r,\dot{r})$ plane or in the $(C^2/r, C\dot{r})$ equivalent one, when the primaries are at their apocenter, $u = (2k + 1)\pi$ and only for symmetric orbits $\dot{r}_o=0$. Results are plotted on the (C, r_o) plane of initial conditions.

. We observe a family of 'Central Periodic Orbits' continuing the circular solutions of the above approximations. This family is stable except between C=1.26 and 1.67 (appearance of 2-periodic orbits).

. Along this family m-periodic families appear. The lower branch of the families and the upper one of the odd families are stable up to a certain distance of the central periodic orbits.

. Along those families 2m-periodic, 3m-periodic families, etc... appear. Such a subdivision 'ad infinitum' is characteristic of non-integrable systems.

. We observe also many 'irregular' families of periodic orbits completely disconnected from the first ones, for instance, in the region of L_4. They are very interesting because of resonance. All seem to be stable.

From the examination of surfaces of section and the systematic research of regions of escape and regions of islands (quasi-periodic motions) we can distinguish between the following cases:

(1) "Very rapid escapes" all under the limit of bounded orbits of the 2 fixed centers approximation.

(2) "Erratic escapes" after one or many rotations around the central periodic orbit.

(3) "Semi-ergodic" orbits.

(4) "Quasi-periodic" orbits, i.e., invariant curves.

(5) "Periodic orbits".

Between types (2) and (3) we can expect "oscillatory motions".

Another interesting problem is to explain analytically the re-appearance of invariant curves (and their pattern) around the central periodic orbit for C < 1.26.

REFERENCES.

Contopoulos, G. (1971), Astron. J. 76, p. 147.

Contopoulos, G. (1970a), Astron. J. 75, p. 96.

Contopoulos, G. (1970b), Astron. J. 75, p. 108.

Hénon, M. (1966), Bull. Astron., (3), 1, fasc. 2, p. 49.

Hénon, M. (1966), Bull. Astron., (3), 1, fasc. 1, p. 57.

Hénon, M. (1965), Ann. Astrophys. 28, p. 599.

Hénon, M. (1965), Ann. Astrophys. 28, p. 992.

ANALYTICAL CONSTRUCTION OF UNSTABLE PERIODIC ORBITS ABOUT THE COLLINEAR POINTS.

D. L. Richardson

Department of Engineering Science, University of
Cincinnati, Ohio

ABSTRACT. A third-order analytical solution for periodic motion about the L_1 and L_2 collinear points of the circular-restricted problem is presented. The three-dimensional equations of motion are obtained from a general Lagrangian approach to the problem. The solution is developed using a technique similar to the classical Lindstedt-Poincaré method.

The complete paper is submitted for publication in Celestial Mechanics.

ON THE STABILITY OF AN EQUILIBRIUM SOLUTION IN HAMILTONIAN SYSTEMS OF TWO DEGREES OF FREEDOM.

H. Rüssmann

Fachberreich Mathematik der Universitat, Mainz, Germany

ABSTRACT. A criterion for isoenergetic stability is established, which can be applied to the Lagrange points L_4, L_5 in the restricted problem of three bodies.

Let

$$x_k' = H_{y_k} \quad , \quad y_k' = -H_{x_k} \quad , \quad k = 1,2 \, ,$$

$$H = H(x,y) = H_2(x,y) + H_3(x,y) + \dots \tag{1}$$

be a Hamiltonian system of two degrees of freedom, where the Hamiltonian is a convergent power series in a neighbourhood of the origin $x=y=0$ with real coefficients and beginning with second order terms. Under the condition, C_m:

$$k_1 w_1 + k_2 w_2 \neq 0 \, , \quad |k_1| + |k_2| \leq 2m \, , \quad \text{where}$$

$\pm iw_1$, $\pm iw_2$ are the eigenvalues of the linearized system, there is a canonical power series transformation

$$(\xi,\eta) \longrightarrow (x,y) = (\xi,\eta) + \dots,$$

such that (1) obtains the Birkhoff normal form up to terms of order 2m:

$$H = H[x(\xi, \eta), y(\xi, \eta)] = F^{(m)}(r) + O_{2m+1} \quad ,$$

$$F^{(m)}(r) = F_1(r) + F_2(r) + \ldots + F_m(r) \quad ,$$

$$F_1(r) = w_1 r_1 + w_2 r_2 \quad , \quad r_k = \frac{1}{2}(\xi_k^2 + \eta_k^2) \quad , \quad k=1,2 \quad ,$$

where $F_1(r)$ denotes the homogenous part of degree 1 in r_1, r_2 of the normal form and O_{2m+1} is the remainder beginning with terms of order $2m+1$.

According to Arnold (1961) the equilibrium solution of (1) is stable in the sense of Liapounov for $t \to \infty$ and $t \to -\infty$ if

$$F_1(w_2, -w_1) = \begin{cases} = 0, & 1 = 1, 2, \ldots, m-1 \\ \neq 0, & 1 = m \geq 2 \end{cases} \tag{2}$$

For $m=2$ this condition reduces to

$$Aw_2^2 - 2Bw_1w_2 + Cw_1^2 \neq 0$$

if $F_2(r) = Ar_1^2 + 2Br_1r_2 + Cr_2^2$. In this case the condition is violated, that is if we have

$$Aw^2 - 2Bw + C = 0 \quad , \quad w = w_2/w_1 \quad , \tag{3}$$

then w is an algebraic irrationality, since A, B, C are rational functions of w over th field of rational numbers and no factor linear in w with rational coefficients can be split away. This happens, for instance, in the restricted problem of three bodies near the Lagrangian points L_4 and L_5. The condiitons C_m are valid for all $m \geq 2$, and consequently, the full Birkhoff normal form exists. One may look if condition (2) is valid for $m=3$ in order to answer the question of stability. But the computation of the Birkhoff normal form is very cumbersome even up to order four, (Deprit and Deprit, 1967) for the Lagrangian points L_4, L_5. So a theorem would be useful stating that algebraic irrationality of w is sufficient for stability. We were not able to prove this in full. But we can prove the following:

Theorem: If $w = w_2/w_1$ is an algebraic irrationality, then the equilibrium solution $x=y=0$ of (1) is isoenergetic stable

(= stable on the energy surface H = 0) in the sense of Liapounov for $t \rightarrow -\infty$ and $t \rightarrow \infty$.

In the restricted problem of three bodies near the Lagrangian points L_4, L_5 the condition (3) is equivalent to

$$D = 644p^2 - 541p + 36 = 0 ,$$

where

$$p = (w + w^{-1})^{-2} = \frac{27}{4} \mu_o (1 - \mu_o) ,$$

$$\mu_o = 0.01091 \ldots$$

as shown in Deprit and Deprit, 1967. If w were rational for p were rational in contradiction to the fact that the polynomial D is irreducible. According to the criterion of Eisenstein, we have 541 is prime, 644 ≠ 0 (mod 541), 36 ≠ 0 (mod 541^2). So w is an algebraic irrationality, and isoenergetic stability of the equilibrium solution x=y=0 is established for $\mu = \mu_o$. The condition of w to be an algebraic number in the theorem can be replaced by a set of inequalities such that w does not produce divisors which are too small in the Birkhoff normal form.

REFERENCES.

Arnold, V.I. (1961), Dokl. Akad. Nauk, SSSR 137, p. 255, or
     Soviet Math 2, p. 247.

Deprit, A. and Deprit-Bartholomé (1967), Astron. J. 72, p. 173.

REGULARIZATION OF TRIPLE COLLISIONS IN THE GENERAL THREE-BODY
PROBLEM

C. Simó

Seccio de Matematiques, Universitat Autonoma de
Barcelona, Spain

ABSTRACT. Two ways of regularization are considered: Siegel's
method, extending the solution analytically through zero if the
power series have exponents of odd denominator, and Easton's
method, i.e., if excluding the collision orbits there is contin-
uity with respect to the initial conditions (in the space of
positions).

Normalized masses may be used with barycentric coordinates
of points of a triangle T.

In the first case one proves that the set of masses for
which triple collision is regularizable is a countable dense
set of curves in T, both for the Eulerian and Lagrangian collisions.

In the case of Easton's or geometric regularization, mean-
ingful both in the physical and computational senses, we begin
with a blow-up of singularity and scaling of time, following
McGehee. One obtains an invariant (triple collision) manifold
(t.c.m.) over which the flow is of gradient type. Rest points
belong to homographic solutions and appear in pairs of collision-
ejection type. Siegel's exponents are (related to) the eigenvalues at such points.

We begin with the collinear case. A necessary condition
for regularization is the coincidence of the unstable manifold
of the collision point with the stable one of the ejection point.
The set of masses for which this happens is the union of a
countable discrete set of curves and a countable set of points
in T. The analysis of the variation of the moment of inertia

along the two branches of such manifolds allows us to establish
that geometric regularization is possible exactly for a countable set of points in T, which may be obtained numerically. All
values given correspond to symmetric configurations of the masses.

   In the planar case the t.c.m. is essentially 5-dimensional.
There are 10 critical points. The dimensionality of stable
(unstable) manifolds at the collisions are 1(4) for Euler's
case and 2(3) for Lagrange. Exchange Dimensions are to be ex-
changed at the ejections. Regularization implies that the
tangent space of the unstable manifold at collision is carried
to the tangent space of the stable manifold at ejection. This
must be verified numerically.

INDEX OF NAMES

INDEX OF SUBJECTS